Likelihood Methods in Biology and Ecology

A Modern Approach to Statistics

Likelihood Methods in Biology and Ecology

A Modern Approach to Statistics

Michael Brimacombe

CRC Press
Taylor & Francis Group
Boca Raton London New York

CRC Press is an imprint of the
Taylor & Francis Group, an **informa** business

CRC Press
Taylor & Francis Group
6000 Broken Sound Parkway NW, Suite 300
Boca Raton, FL 33487-2742

Printed on acid-free paper
Version Date: 20181108

International Standard Book Number-13: 978-1-5848-8788-1 (Hardback)

Visit the Taylor & Francis Web site at
http://www.taylorandfrancis.com

and the CRC Press Web site at
http://www.crcpress.com

Many Thanks to GJH, my wife Barbara and children without whose support, direction, and guidance this work could not have been written.

Contents

Preface

Introduction

This book is written for applied researchers, scientists, consultants, statisticians, and applied scientists. It reflects many years of collaborative work with applied scientists and biomedical researchers. It provides an accessible handbook of available statistical methods for scientific settings where there is an assumed theoretical model that can be represented using a likelihood function. This is then viewed from various perspectives to draw inferences and conclusions. Most standard statistical methods fall in this category. The book also reviews and discusses standard methods of data analysis, model selection, and statistical analysis, and how to apply and interpret them in real-world situations. The mathematics is present, but not overwhelming.

This book assumes a basic course on probability and statistics and a second course on linear models or regression, as well as some exposure to data analysis. It is focused in terms of applications on the areas of ecology and biology, both very large fields, but the methods here can be applied to a wide variety of research areas. The mathematics are kept to a minimum and concepts are introduced in Parts I–III and applied in Part IV. The book also discusses some of the pitfalls that can be encountered when designing experiments and analyzing data, including the need to be very careful when analyzing large datasets that are not pre-designed for statistical analysis. The case studies in Part IV are written according to a template that may be useful in a variety of applied fields. Some of the theoretical sections have been components of courses taught at various institutions. The applied sections reflect aspects of various published collaborative scientific projects to which the author has contributed. The datasets are based on these studies but related simulations are used for the purpose of this book.

It is difficult to find books that aim at younger researchers (including statisticians and mathematicians dealing with real-world data challenges) and researchers whose background is not mathematics and provide them with (i) an overview of the statistical and data analytic tools available for analyzing data and developing statistical models and (ii) a basic template for applying them in real-world problems where the clarity of mathematical assumptions gives way to the often difficult terrain of real-world science and modeling. The

Likelihood Methods in Biology and Ecology

statistical methods and their justifications and limitations are clearly discussed with the mathematics present, but again not overwhelming the presentation.

This book can be used for a course in applied statistics and for statistical courses taught in the applied sciences, including biology and ecology, but also anywhere the underlying principles and intuition of statistical and data analytic techniques are being introduced. After going through the material in this book the student should have a good basic understanding of the statistical methods available when using parametric statistical models, linear and generalized linear models. The relevance of the likelihood function to both the frequentist and Bayesian perspectives should be clear. The student should also have a good idea of how these two approaches can be viewed as complimenting each other, providing useful sets of analytic tools to better understand the outcomes of the experiment and the conclusions supported by the model-data combination and the assumptions in applying each perspective.

Overview

The use of statistical methods to analyze and understand data has been ongoing now for over 100 years. For the past 60 years the computer has been an aide in this process. While mathematical models often dominate the application of statistics, the underlying scientific context is essential, as is the analysis of the observed dataset. The specific approaches taken to the collection of data, the definition of variables and selection of mathematical models should reflect the understanding and context of the scientific question being studied. These change over time as technologies improve, as do the accepted methods for the statistical analysis of model-data combinations.

The importance of the likelihood function in statistical theory and applications is emphasized in this book. It links the assumed theoretical model and the observed data. It guides the selection of statistics, provides a context for parameter estimation and the testing of specific hypotheses, and allows for the measuring of information. Bayesian and frequentist methods both use the likelihood function and the information parsed within. Depending on what we assume, they provide differing but related insights. This is examined here both through review of basic methodology and also the integrated use of these approaches in particular case studies.

The study of these two approaches to analyzing the information in the likelihood function is often taught separately, but they can be taught together as they represent different perspectives on the same basic likelihood and data combination. The frequentist perspective is mostly pre-experimental with probability describing possible outcomes and the likelihood function providing both a useful descriptive summary and tool for estimation and testing. The Bayesian perspective is conditional on the observed data, and through use of the likelihood function, primarily updates existing beliefs. It uses probability as a scale to describe relative degrees of belief. This can then be extended to additional inferential measures.

The application of statistical methods to observed data is examined here in the context of examples drawn from biology and ecology, two wide ranging areas of research. Both fields have long histories of using frequentist statistics and both have recently seen a rise in the application of Bayesian methodology. From ecology (i) a study in species abundance and (ii) a study in soil erosion and land usage. From biology (i) immunity and dose response in relation to aquaculture, (ii) patterns of genetic expression in mouse liver cancer and (iii) antibiotic resistance in relation to genetic patterns in tuberculosis. The examples chosen reflect actual research and related published results. The datasets utilized here are simulated versions based on actual datasets.

Somewhat uniquely the focus on the likelihood function here is broadened into an overview including both frequentist and Bayesian approaches to statistics. They are seen as complementary, reflecting different perspectives in relation to probability and how the information in the likelihood function is to be interpreted. This allows many standard approaches to be presented in a single book. Many statistical books from both the frequentist and Bayesian perspectives choose to emphasize differences, but in many basic experimental settings they are very similar at their core which is the likelihood function. A unique integrated approach is taken to their application to the analysis of data sets here. The methods do not contradict each each other, with the insights they provide in applied settings reflecting the assumptions made.

The template offered here to help guide the formal writing up of analyses focuses on the scientific context, the initial study design and data collection, initial data analysis, the assumptions and justifications of specific models, the fitting of these models to the observed data and the description of the experimental results through the fitted likelihood function. Both frequentist and Bayesian inferential summaries are relevant to parameter estimation and hypothesis testing. These are then presented as an integrated summary of the results in relation to the scientific setting.

There is much history and a wide ranging set of references reflecting the broad application of statistical and data analytical methods. R.A. Fisher (geneticist and statistician) studied the linked concepts of probability and information while developing the theory of population genetics and the application of statistics to areas such as agriculture, biology, ecology and aspects of genetics, among others. In relation to theory, his work includes the nature of probability, sufficiency, likelihood, maximum likelihood estimation, randomization tests, Fisher information, ANOVA and least squares, categorical data analysis, the analysis of variance (ANOVA), chain binomial models in epidemiology, the design of experiments especially incorporating replication, and the use of multivariate approaches in the analysis of data. His work balanced the mathematics necessary in the use of statistical methods with scientific relevance and context. He was not a fan of Bayesian ideas, but his likelihood function, especially the score function, lies at the heart of how Bayesian methods process experimental information.

The use of randomization and permutation tests and re-randomization methods of observed data are not emphasized in this book. These methods have conceptual challenges related to the nature of probability in relation to observed data values. Also if they are based on statistics that are not sufficient or optimal they violate the sufficiency principle (the use of minimal sufficent statistics whenever possible) and by re-randomizing they risk throwing off beneficial effects of initial randomization performed as part of the experiment or study. That said these methods can be seen as useful data analytic methods that are often used to investigate data based properties of complex structures. They include modern regression methods and the fitting of network models.

Computing is not formally presented here. Many good introductory books exist in regard to the R program and environment or for example the SAS package or even "point-and-click" programs such as Minitab and versions of SAS or SPSS. Bayesian calculations remain a slight challenge, but there are many accessible Bayes packages in the R environment or a package such as WinBugs that provide access to standard Bayes summary and inferential measures.

Using the Book

This book is comprised of four parts containing twelve chapters. Part I is comprised of Chapter 1 focusing on statistical models in scientific research, a broad overview of the place of statistics in science along with some questions to ask when planning the design and analysis of a study or experiment. Some of the basic statistical and modeling tools available are introduced including initial discussion of frequentist likelihood and Bayesian likelihood methods.

Part II is entitled, "Basic Tools for Data Analysis, Study Design and Model Development" and contains three chapters along with exercises and references for the entire section at the end of Chapter 4. The first chapter examines basic methods of data analysis including graphical methods and principal component analysis. The second reviews basic issues in the design of experiments and includes issues such as the need for replication. Simpson's paradox is also discussed in terms of trying to understand and model patterns and associations among variables in the data and the potential effects of aggregating data across sub-groups. The use of mis-specified models, mathematical models that are incorrect in their basic structure and may incur bias in results, are also briefly addressed. Data analysis is not a replacement for understanding the basic scientific structures among the variables being studied. Chapter 4 discusses how to develop probability distributions to model prior beliefs in relation to basic statistical models and structures in the data.

Part III is entitled, "Likelihood Based Statistical Theory and Methods: Frequentist and Bayesian." Chapter 5 discusses frequentist likelihood based theory and some basic models. Chapter 6 discusses likelihood based Bayesian methods and some of the standard approaches to developing, using and reporting information from this perspective. Exercises for both chapters are at the end of the section.

Part IV addresses how to apply the methods and perspectives discussed in Parts I–III, in the context of applications drawn from biology and ecology. A common template is used in the development and reporting of results from an overall integrated Frequentist-Bayesian applied perspective. For each example, the overall patterns and inferential summaries observed are discussed and used to develop insights into the scientific issues addressed. Each case study has some additional exercises and suggested further analyses.

Statistical science departs from mathematics in that it must deal with real world data and come to some conclusions, typically up to a given level of accuracy. It departs from data analysis in that it acknowledges and applies mathematical summaries both to previously existing knowledge and in the development of the parametric models that yield the likelihood based inference it provides. It is a challenging field with many possible approaches and issues to consider. Hopefully this book gives the reader a practical introduction to its methodology, uses, and applications.

My sincere thanks to CRC Press, in particular David Grubbs and Shashi Kumar.

M. Brimacombe Ph.D.

Part I

Introduction

1

Statistical Models in Scientific Research

1.1   Statistics in Science

The history and use of probability and statistical methods in the sciences is a long and varied one. The birth of modern statistics in the early 1900s was primarily the work of scientists active in fields such as population genetics, agriculture, biology, astronomy and geology. They played major roles in developing the statistical methodologies and approaches that scientists now routinely use across a wide variety of fields. Statistics, as an applied science, sits on two sets of foundations. One is primarily mathematical, related to the core set of sampling ideas and related probability formalisms that one can find taught in most mathematical statistics courses, based on frequentist ideas, with extensions to the learning model format of the Bayesian statistical approach. The other is scientific, the form of which is specific to each area of science where the level of observed randomness in the empirical data is significant enough to warrant modelling and/or application of judgements expressed on a probability scale.

While the incorporation of randomness was present in the work of Gauss, Laplace and the early probabilists in the late 1800s (Stigler, 1986), much of the initial impetus for the development of modern statistical methods was due to the large role played by naturally occurring variation in population genetics, an area where variation for the sake of variation is of key interest. Genes that vary on some scale from generation to generation may play a role in the onset of future population changes, as they are available to react to altered selection pressures in the environment.

R.A. Fisher, scientist and population geneticist, often viewed as the father of modern statistics (Box 1978, Efron 1998), extended the statistical approaches and methods that he initially developed in the setting of population genetics to the areas of biology, agriculture, geology and many experimental sciences. In doing this he developed and refined much of the basic mathematical framework of modern statistics, often laying out the analysis in very interpretable constructs, developing tools that allowed scientists to think through statistical aspects of the model and data in a useful manner (Box, 1978). The need to incorporate some level of randomness in most applied scientific settings has given these statistical methods wide application, especially through the use of randomized designs, ANOVA and least squares based analysis, multivariate analysis, the likelihood function and the related concept of Fisher information.

The application and interpretation of statistics over the years has also lead to a long history of discussion and argument over several related concepts; the nature of probability, the role of experimental design and pre-planning in observational or experimental studies, the appropriate degree of objectivity in the interpretation of statistical results, the use of personal belief and judgement as the basis of a probabilistic assessment. Approaches to these topics in relation to statistical modeling are still being discussed and debated (Efron, 1998).

By focusing on the need to interpret the likelihood function and the information it provides, we will see that the two main statistical approaches, Bayesian and frequentist, distinct in the perspectives they employ, can be integrated into practical analyses of many scientific problems. Recent advances in the computing necessary for the application of Bayesian methods allows for practical application and comparison of both perspectives in real-world settings. This is further discussed in Parts II and III and applied in Part IV.

1.1.1   Guidelines to Statistical Model Building

First we discuss how to approach basic model building with simple statistical concepts. There are few areas of endeavor as broadly defined and applied as "statistics." Theoretically, in application and in computation, the name "statistics" can invoke many different levels of understanding and interpretation. In the physical, life and social sciences, statistics is used to help organize, collect and model characteristics and potential associations of interest where measurements are subject to random fluctuation.

While there are common themes and approaches to the doing of applied statistics, they tend to be subjugated to the goals of the scientific discipline of interest. To be of real use, statisticians and researchers applying statistical methods should understand the relevant literatures underlying the scientific interpretation of the models they are helping to develop. Indeed empirical statistical modelling, often within the framework of a parametric model reflecting scientific theories, may be the main tool for the analysis and comprehension of the data that is being collected, especially in large, complicated datasets.

1.1.2   Questions and Answers

In applications, there is always a context. This context is created or underpinned by the scientific literature, practical knowledge and goals of the specific discipline in question. Before any statistically related application or approach in the specific field can be developed, some simple, yet crucial questions need to be considered.

These typically include:

Question Set 1

- What variables or characteristics should be measured?

- What variables or characteristics can be measured in a meaningful manner?

- What are the overall goals of the experiment or study?

- What are the central hypotheses to be investigated?

- Which potential relationships between variables are of interest?

- What time frame should be employed?

- On what scale(s) should response(s) and other variables be assessed? Can the data be aggregated/manipulated onto interpretable indices that simplify the analysis?

- At what point in the scientific development does random error or fluctuation become relevant and in need of formal modelling?

- How do observed correlations among responses or between explanatory variables affect the determination of scales of measurement and variables of interest?

- Is there a willingness on the part of the researcher to accept discretization of certain continuous variables?

- Which parametric models are available that will allow for these goals to be examined and the data summarized in the most straightforward and scientifically relevant fashion?

Once the above questions are answered, enough of a context will be defined to allow for a second set of questions which relate to the modeling of stochastic aspects of the analysis. These typically include:

Question Set 2

- What is the statistical design and/or randomization pattern to be used in collecting the data?

- What are the potential sources of bias?

- Are there variables confounding our interpretation of results?

- What is the assumed model and resulting likelihood function?

- Are there random effects or nested/hierarchical structures that can be placed in the model?

- If a likelihood can be defined, should a Bayesian or frequentist approach be employed to examine the data? Can both give insight?

- Do prior beliefs exist in relation to the populations parameters of interest? How would we model these?

- What are the hypotheses (in parametric form) to be tested?

- Which specific tests ("nonparametric" versus "parametric") should be employed and in what sequence?

- What is the statistical power of the test(s) involved?

- At what sample sizes are p-values potentially misleading or not clinically/scientifically useful?

- How stable is the resulting statistical analysis?

In many areas of research Question Set 2 takes on a precise, practical meaning only once the initial scientific context determined by the answers to Question Set 1 are obtained. The basic questions of hypotheses, time frames, variable selection, scale of measurement and available resources must first be addressed.

There have been attempts to formalize the process of practical scientific investigation and information collection as it underlies the initial development of context, models and beliefs to follow. Some of this literature has the formal title measurement theory (Hand, 1996). This literature discusses the need to define basic contexts, variables and measurement tools before applying statistical tools. In this book we focus on examples drawn from the fields of ecology and biology, which have long and detailed histories and literatures addressing model development and statistical application.

Awareness of the need to more completely formalize how we learn in relation to existing beliefs finds a natural home in well designed Bayesian statistical settings where the interaction of previously existing beliefs with the modeling process is formally acknowledged and modeled. In the frequentist setting there tends to be a discomfort with this, as it seems to open the door to non-objective elements in the subsequent analysis. Frequentist statistics often reflects an older more conservative experimentalist approach to scientific discovery. In this setting we test hypotheses via contradiction and focus on individual experiments and results developed in an objective context. As we shall see, objectivity is always contextual, and a degree of subjectivity, as long as it is carefully applied and acknowledged, can be a useful approach in developing helpful tools and perspectives. The fact that all researchers must subjectively choose a statistical model within a pre-defined scientific context is an important point that should not be forgotten. In Figure 1 an overview of the analytic process relevant to Question Set 2 is given.

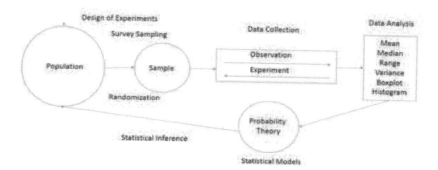

FIGURE 1.1
The Elements of Inference

1.1.3 Basic Statistical Models

Statistical models tend to have two basic elements; a core mathematical model, typically of the form $y_i = \beta_0 + \beta_1 x_{1i} + \ldots + \beta_p x_{pi}$ relating a response y and a set of explanatory variables $x_1, \ldots, x_p$ for each set of individual measurements $i = 1, \ldots, n$. They also have independent random components (ε_i), usually additive in nature, having a probability distribution $f(\cdot)$, with the two often being written together as

$$y_i = \beta_0 + \beta_1 x_{1i} + \ldots + \beta_p x_{pi} + \varepsilon_i \tag{1.1}$$
$$\varepsilon_i \sim f(\cdot) \tag{1.2}$$

This is a linear (regression) model, often the simplest model to use when relating variables. Other statistical models having an additive random component may also have more realistic nonlinear forms, such as

$$y_i = \beta_0(1 + e^{-\beta_1 x_i}) + \varepsilon_i, \tag{1.3}$$

which reflects a specific shape. These nonlinear regression models are more typical of those applied in biology or ecology. Figure 2 shows the BOD dataset with the above nonlinear model fit to the data.

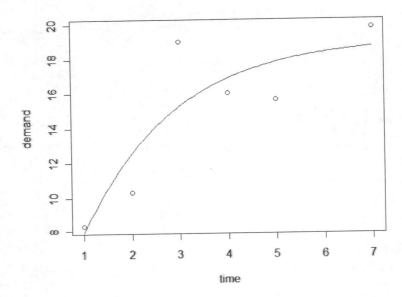

FIGURE 1.2
Nonlinear Regression Model and Data

Some models require rescaling to achieve these simple model forms. Basic transformations such as square roots, logarithms, squares or odds ratios tend to be employed. For example, a log transformation of the response y in (1.1) gives us the model:

$$\log y_i = \beta_0 + \beta_1 x_{1i} + \ldots + \beta_p x_{pi} + \varepsilon_i.$$

It is worth noting that the meaning of the parameters themselves may alter when such transformations are applied. For example, the parameter β_1 goes from being a measure of the rate of change in y for a small change in x_1 (holding other variables constant), to the percent change in y for a small change in x_1 once y is log-transformed.

In this book many types of statistical models are discussed and applied. The choice of statistical model tends to reflect both the underlying scientific context and statistical interpretation. The flexibility available to statisticians, ecologists, biologists and others in developing mathematical models is both impressive and fraught with challenges.

Not all mathematical models lend themselves easily to statistical analysis. Nonlinear models and other complex models in mathematical biology and ecology that use multiple differential equations in their definition may not allow for simplification and use of standard statistical technique (see for example Seber and Wild, 1989). As well, recent advances in the application of nonlinear dynamic models resulting in chaotic behaviors can create modeling difficulties such as unstable liklihood functions (Givens et al., 2002). These limitations are discussed in the text and exercises and should be a consideration in the selection and application of statistical models. There is often a trade-off to be made between a useful model and a more realistic one.

1.1.4 Likelihood Function

The most direct way to approach Bayesian statistical models and most frequentist parametric statistical models is through the well-known likelihood function developed by R.A. Fisher (1922). This approach has had wide applicability to many settings and is the core statistical concept in most applications allowing for a parametric statistical model. When available, the likelihood function, which can have its difficuties (Bahadur, 1958), is the common, practical starting point for application of either frequentist or Bayesian methods.

To begin, assume we have a candidate probability distribution describing our response variable y, call it $p(y_i; \theta)$, where y_i is the observed value for the ith individual and θ is a generic parameter, often taken to represent a population characteristic of interest. For ease of exposition we assume independent observations and we can write:

$$p(y_1; \theta) \cdot p(y_2; \theta) \cdots p(y_n; \theta) = \Pi_{i=1}^n p(y_i; \theta)$$

as the probability our observed sample will occur. If we take the responses $y_1, \ldots, y_n$ as being observed and we plug their actual values into this function, we can view it as a function of θ alone and write the resulting likelihood function as:

$$L(\theta|\, data) = c \cdot \Pi_{i=1}^n p(y_i; \theta) \qquad (1.4)$$

where c is an arbitrary constant. Note that, as a function of θ, $L(\theta|\, data)$ is (i) not a probability distribution and (ii) is informative regarding which values for θ are supported by the data on a relative scale. Note that in some Bayesian settings, the arbitrary constant is not viewed as part of the definition.

1.1.5 Frequentist Interpretation

Standard frequentist based statistical methods, where there exists a clearly defined likelihood, depend directly on properties of the logarithm of $L(\theta|\,data)$, which serve as the basic building block for statistical inference. Properties of the likelihood function are the basis of several commonly used test statistics. The maximum of the function defines the maximum likelihood estimator $\hat{\theta}$, and the difference $(L(\theta_0|\,data) - L(\hat{\theta}|\,data))$, on a log scale, is the basis of the likelihood ratio statistic. Comparison of likelihood based derivatives underlies the score function. Frequentist sampling theory is applied to find the distribution of these quantities and derive 95% confidence intervals and p-values. These approaches are both optimal and practical much of the time and are described in more detail in Part 3.

In large sample statistical applications the expected local curvature of the log-likelihood function about its mode, the Fisher information, provides an approximation to the underlying variation in the model. This has wide application in biological and ecological systems and many other applications with relevance to both Bayesian and frequentist inferential methods. The Fisher information is directly related to the Cramer-Rao lower bound which provides a large sample estimate of variation in many settings. This is also further discussed in Parts 2 and 3.

Note that when a normal distribution is assumed for $p(y_i; \theta)$, the resulting likelihood function is proportional to a normal density. To see this, assume an independent sample made up of observations drawn independently from a $N(\theta, 1)$ distribution. The likelihood function is given by:

$$L_{normal}(\theta\,|\,data) = c \cdot \exp\{-\frac{n}{2}(\theta - \overline{x})^2\} \tag{1.5}$$

which is a bell curve centered at the sample average $\overline{x}$.

This intriguing fact is the basis for the often occurring agreement between Bayesian and frequentist methods in many standard situations, especially in the presence of large sample sizes when the central limit theorem converges the shape of the likelihood to normality for many initial choices of $p(y_i; \theta)$.

1.2 Bayesian Statistical Analysis and Interpretation

In this book, we apply Bayesian methods, in tandem with frequentist methods, to a variety of applied models in biology and ecology. The application of Bayesian methods to real-world scientific problems has grown dramatically in the past few years. Many settings that have hierarchical structures in the collection and interpretation of data are particularly well suited to the approach.

The approach taken here to Bayesian methods is an applied one, focusing on the subjective interpretation and assessment of likelihood based in-

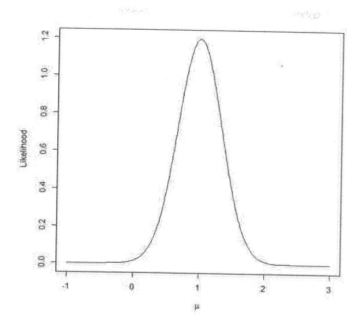

FIGURE 1.3
Normal Likelihood

formation. The Bayesian approach actually underlies early use of statistical methods by Laplace (Stigler, 1986) and was later carefully developed by the geologist Harold Jeffreys. His interaction with R.A. Fisher is interesting in itself (Fisher 1934, Jeffreys 1934, Lane 1980). A very useful general exposition of Bayesian methods remains the book by Box and Tiao (1973), where the overlap and inter-relation with frequentist approaches is addressed. More recent work by Bernardo and Smith (1994) and O'Hagan (1994) can be consulted for overviews of specific aspects of theory. See also Congden (2007) and Gelman et al., (1995).

It is sometimes said that there are as many ways to apply Bayesian methods as there are Bayesian statisticians, but here we take an approach motivated by application and understanding of the importance of the likelihood concept. We focus on the likelihood function as the key aspect of statistical modeling. We define this in detail below. Bayesian methods require a probability density with which to calibrate the likelihood function, called the prior distribution. This modulates the likelihood function itself into a probability density (the posterior density) as a function of the parameters. Some describe this in reverse terms; the model-data combination acting through the likelihood function acts to re-weight existing prior beliefs. Thus the posterior reflects a type of learning behavior, with the prior density being the initial baseline knowledge, the likelihood providing model-data based information and modifying that knowledge and the posterior reflecting the modified result.

Developing prior information to reflect baseline beliefs regarding the population characteristics (parameters) requires a deep understanding of the variables and measurements that define the scientific problem at hand. Often Bayesian methods are viewed with skepticism as they require more assumptions than standard methods. But these assumptions may be justified if proper care is given to understanding the science and context involved. Bayesian methods are best applied in research areas that are fairly well developed, but can also be applied to most standard settings.

In the context of hypothesis testing, a key component of frequentist statistics, Bayesian methods are more formally based in a learning or belief-modifying setting where evidence is slowly collected, assessed and beliefs adjusted to reflect real-world evidence (often answering the question: What have we learned?). This is a very different setting from the frequentist paradigm of an assumed null hypothesis and the testing of this hypothesis essentially through contradiction (answering the question: Can we reject the null hypothesis?). We discuss these differences in perspective as they arise in the various examples in Part IV.

Bayesian statistics can typically be applied anywhere standard statistics can be applied, especially when a likelihood function can be found to describe the underlying processes generating the data. Issues regarding data collection, design, power etc. typically play little formal role in Bayesian analyses. These usually are assumed to have been carried out properly as the observed data is conditioned upon. This has lead to the idea of developing a formal integration of frequentist randomization principles, as they relate to design, and Bayesian statistical models as a potentially integrated theory for modern statistics (Little, 2006). The approach taken here uses both approaches as distinct perspectives, motivated by the need to solve real-world biological and ecological problems.

The basic application of Bayesian methods is actually straightforward and we briefly summarize it here. More details are provided in Part III.

1. Given the scientific problem to be investigated and the selected probability model and resulting likelihood function, define a prior distribution $p(\theta)$ for the parameter(s) θ of the model. This should be carefully elucidated from existing belief and knowledge, though we often assume a flat, non-informative prior or a prior following standard probability density form.

2. Re-calibrate the likelihood function so it is a probability distribution for θ using the assumed prior distribution. The resulting density is

called the Bayesian posterior distribution and is given by;

$$p(\theta|\,data) = k \cdot p(\theta) \cdot L(\theta|\,data)$$

where k is the relevant constant of integration.

3. Integrate out parameters not of interest from the overall posterior density to obtain marginal posterior densities for individual parameters of interest θ_i. In theory the modification of the likelihood function into a probability distribution allows for application of simple probability calculus to obtain the required marginal probabilities for specific population characteristics or parameters of interest. This should in theory simplify most statistical applications, though numerical integration can be challenging.

4. Use the obtained marginal posterior probabilities, $p(\theta_i|\,data)$, to examine specific hypothesis and their related level of support in the given model-data setting or provide central 95% "credible" regions for estimation of θ. In the Bayesian context measures and levels of evidence are based upon such post-experimentally valid "posterior" probability statements. In terms of estimators and their properties, the posterior density inherits many of the good properties of likelihood, especially in relation to sufficiency and information. Posterior odds ratios and Bayes Factors are also available. See Part III.

The challenge in applying Bayesian methods is the justification of the prior distribution $p(\theta)$, which reflects pre-existing beliefs regarding potential values of θ elucidated on a probability scale, and the surprising difficulty of averaging out unwanted parameters or elements of θ in the posterior function in higher dimensions. In return for overcoming these difficulties, the Bayesian approach gives us detailed information for θ as well as a calibrated approach to the assessment of evidence in support of hypotheses. The reject/do not reject approach underlying frequentist statistics may be augmented by observed levels of support that can be more informative than simply comparing p-values to an assumed 0.05 Type I error cutoff. The question becomes both "How much did we learn?" *and* "Can we reject the null hypothesis based on this sample?"

Note again that the prior distribution models personal beliefs regarding the parameter θ. This is then updated by use of the model-data combination reflected in the likelihood function and rescaled as the posterior distribution $p(\theta|\,data)$. Both prior and posterior densities use the formalism of the probability scale to represent preferences and beliefs. Some history in this regard can be found in de Finetti (1931), Savage (1961) and Good (1959). Prior distributions may be rooted in previously observed data, previously observed likelihoods or aspects of the current data itself, giving rise to the empirical Bayes approach. This compromise has been around a while (Carlin and Louis 1996, Efron 2010) and has been applied in environmental models (Ellison, 2004).

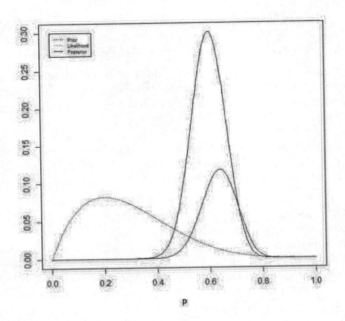

FIGURE 1.4
Beta-Binomial Prior, Likelihood and Posterior

1.3   A Comparative and Practically Integrated Approach

In this book we take the approach that both the Bayesian and frequentist perspectives have useful contributions to make when we attempt to interpret and model real-world scientific problems. From the practical common ground of the initial definition of the scientific problem and the subsequent definition of the likelihood function, both approaches can be used to investigate specific hypotheses, basic modeling and prediction. Distinct in perspective and interpretation. Based on a learning paradigm and perspective, the Bayesian approach is technically an extension of the likelihood based frequentist approach, calibrating the likelihood function into a density on the parameter space Ω by use of a prior density for θ. The quality of the resulting Bayesian analysis depends on the accuracy of the likelihood function and the stability of resulting inferences over a set of potential prior densities. The result is an overlapping but conceptually distinct collection of statistical tools available for scientific analysis.

In Part IV, we present comparatively and practically integrated analyses of various ecological and biological problems. In each problem the science is reviewed; the context, variables and key questions to be addressed. The statistical model and resulting likelihood function to be applied is then reviewed, with both frequentist and Bayesian perspectives on the model employed and results obtained with discussion. The goal throughout is to examine how hypothesized mathematical models may be fit to real data and interpreted, both from a frequentist-likelihood and a Bayesian perspective, developing an understanding of how these approaches augment each other in the modeling process.

In the setting of Biology in Part IV we will examine the following topics and problems:

A. Toxicology: Dose Response in Aquaculture Trials. (Application: ANOVA, ANCOVA, replication).

B. Genetic Associations: Patterns of Genetic Expression in Mouse Cancer. (Application: categorical data, correlations, genetics, high dimensional databases).

C. Antibiotic Resistance: Modeling Antibiotic Resistance in Relation to Genetic Patterns in Resistant Tuberculosis. (Application: logistic regression, cluster analysis).

In the setting of Ecology in Part IV we will examine the following topics and problems:

A. Biodiversity: Modeling Species Abundance in Relation to Ecological Patterns. (Application: linear and nonlinear models, cluster analysis, power law).

B. Soil Erosion: Erosion Patterns in Relation to Season and Land Usage. (Application: cluster analysis, loess, linear models, ANOVA).

1.4   Computing

The computing here is carried out using standard packages for data and frequentist statistical analyses. These include R and Minitab. Most Bayesian statistical analyses are carried out using R and the freely available WinBugs package which uses various MCMC based approaches to obtain required multiple integrations. The WinBugs site is http://www.mrc-bsu.cam.ac.uk/bugs/.

Computing in real-world problems can be a serious challenge and there are typically several approaches to the computation of any problem. This book is not a primer on statistical computing, but much of the presented analysis and comments will help guide the researcher seeking to apply the approaches on their own.

1.5 Bibliography

[1] Bahadur R.R. (1958). Examples of Inconsistency of Maximum Likelihood Estimates. *Sankh'ya* 20, p. 207–210.

[2] Bernardo J.M. and Smith A.F.M. (1994). *Bayesian Theory.* John Wiley and Sons Inc. New York, NY.

[3] Box J.F. (1978). *R.A. Fisher: The Life of a Scientist.* John Wiley and Sons Inc.

[4] Box G.E.P. and Tiao G.C. (1973). *Bayesian Inference in Statistical Analysis.* Addison-Wesley Publishing Company Inc. Reading, Massachusetts.

[5] Carlin B.P. and Louis T.A. (1996). *Bayes and Empirical Bayes Methods for Data Analysis.* Chapman & Hall, New York, NY.

[6] Congden P. (2007). *Bayesian Statistical Modelling*, 2nd Edition, Wiley.

[7] de Finetti, B. (1931). Funzione caratteristica di un fenomeno aleatorio. *Atti della R. Academia Nazionale dei Lincei, Serie 6. Memorie, Classe di Scienze Fisiche, Mathematice e Naturale*, 4:251–299.

[8] Efron B. (1998). R.A. Fisher in the 21st Century. *Statistical Science* 11, p. 95–122.

[9] Efron B. (2010). *Large Scale Inference: Empirical Bayes Methods for Estimation, Testing and Prediction.* IMS Monographs, Cambridge University Press.

[10] Ellison A.M. (2004). Bayesian Inference for Ecology. *Ecology Letters* 7, p. 509–520.

[11] Fisher R.A. (1922). On the Mathematical Foundations of Theoretical Statistics. *Philosophical Transactions of the Royal Society* A, 222, p. 309–368.

[12] Fisher R.A. (1934). Probability, Likelihood and the Quantity of Information in the Logic of Uncertain Inference, *Proceedings of the Royal Society* A, 146, p. 1–8.

[13] Gelman A., Carlin J.B., Stern H.S., Rubin D.B. (1995). *Bayesian Data Analysis.* Chapman & Hall, New York, NY.

[14] Givens G. and Poole D. (2002). Problematic Likelihood Functions from Sensible Population Dynamics Models: A Case Study, *Ecological Modeling* 151, p. 109–124.

[15] Good I.J. (1959). Kinds of Probability, *Science*, Vol. 129, p. 443–446.

[16] Hand D.J. (1996). Statistics and the Theory of Measurement. *J. Royal Statist Soc. Ser A* (Statistics in Society), Vol. 159, No. 3, p. 445–492.

[17] Jeffreys H. (1934). Probability and Scientific Method, *Proceedings of the Royal Society* A, 146, p. 9–16.

[18] Lane D.A. (1980). Fisher, Jeffreys and the Nature of Probability, p. 148–160, in R.A. Fisher: An Appreciation. Fienberg S.E. and Hinkley D.V. (eds). *Lecture Notes in Statistics, Vol 1*, Springer-Verlag, New York.

[19] Little R. J. (2006). Calibrated Bayes: A Bayes/Frequentist Roadmap. *J. Am. Statist. Assoc.* 60, p. 213–223.

[20] O'Hagan A (1994). *Kendall's Advanced Theory of Statistics*, Volume 2B, Bayesian Inference. John Wiley and Sons, New York.

[21] Savage L.J. (1961). The Foundations of Statistics Reconsidered. *Proceedings of the Fourth Berkeley Symposium on Mathematical Statistics and Probability*, Berkeley and Los Angeles, University of California Press, 1961, Vol. 1, p. 575–586.

[22] Seber G.A.F. and Wild C.J. (1989). *Nonlinear Regression.* John Wiley, New York.

[23] Stigler S.M. (1986). *The History of Statistics.* Belknap Press, Harvard.

1.6   Suggested Readings

1. Baker C.S., Clapham P.J. (2004). Modelling the Past and Future of Whales and Whaling. *TRENDS in Ecology and Evolution* 19, p. 365–371.

2. Bernardo J.M. and Smith A.F.M. (1994). *Bayesian Theory.* John Wiley and Sons Inc. New York, NY.

3. Box G.E.P. and Tiao G.C. (1973). *Bayesian Inference in Statistical Analysis.* Addison-Wesley Publishing Company Inc. Reading, Massachusetts.

4. Edwards A.W.F. (1992). *Likelihood.* The Johns Hopkins University Press. Baltimore, Maryland.

5. Fienberg S.E. and Hinkley D.V. (eds). (1980). *R.A. Fisher: An Appreciation.* Lecture Notes in Statistics, Vol 1, Springer-Verlag, New York.

6. Frigessi A., Holden M., Marshall C., Viljugrein H., Stenseth N.C., Holden L., Ageyev V., Klassovskiy N. (2005). Bayesian Population Dynamics of Interacting Species: Great Gerbils and Fleas in Kazakhstan. *Biometrics* 61, p. 230–238.

7. Gelman A., Carlin J.B., Stern H.S., Rubin D.B. (1995). *Bayesian Data Analysis.* Chapman & Hall, New York, NY.

8. Gilks W.R., Richardson S., and Spiegelhalter D.J. (1996). *Markov Chain Monte Carlo in Practice.* Chapman & Hall, New York.

9. Lange N., Ryan L., Billard L., Brillinger D., Conquest L., Greenhouse J. (1994). *Case Studies in Biometry.* John Wiley and Sons Inc. New York, NY.

10. Luo R., Hipp A.L., Larget B. (2007). A Bayesian Model of AFLP Marker Evolution and Phylogenetic Inference. *Stati. Appl. Genet. Mol. Biol.* 2007 Jul; 88(7):1813–23.

Part II

Basic Tools for Data Analysis, Study Design and Model Development

2

Data Analysis and Patterns

2.1 Data Analysis, Beliefs, and Statistical Models

Statistical models, both the mathematical relationships among variables and the probability distributions within which they are embedded, need the support of initial data analysis and/or theoretical support to be useful. This is true in both frequentist and Bayesian contexts. All applications of models should be based on a clear understanding of the data and the scientific context within which the data is collected. This includes the ranges and appropriate scalings of the variables to be measured, typical or expected patterns in the data, potential interactions among the variables, expected differences in patterns across different strata and the demographics of the subjects. The conceptual paradigms or sets of questions that were introduced in Chapter 1 are also available to be employed in the consideration of a new experiment or study.

Here in Chapter 2 we introduce the basic materials and data analytic tools of statistical and probabilistic analysis. We review standard graphical methods to investigate the properties of data. This is an important component of both model justification and the development of prior belief. The development and formulation of prior beliefs is discussed from various perspectives, a discussion which continues in Chapter 6. We also examine the standard set of parametric statistical models underlying the choice of likelihood functions typically available to applied researchers. These models are fundamental tools for scientists applying likelihood based statistical methods, relevant to both frequentist and Bayesian perspectives.

2.2 Basic Graphical and Visualization Tools

When seeking to model real-world phenomena, especially in biological or ecological settings where nonlinearity and the rescaling of measurements are common, visualization and graphical representation are important tools. These include scatterplots, boxplots, histograms and multivariate versions of these when many variables are to be examined. More complicated patterns in the data may be investigated by use of correlation matrices, principal component decompositions, cluster analysis, or conditional plots where levels of a third variable (or more) are held constant. Many computational packages are now available that provide excellent graphics. R (www.r-cran.com) is such a package. A simpler package providing plots and data analytic approaches also applied in this chapter is the Minitab package (www.minitab.com). We will not review a general philosophy of how to present data and how various plots can mislead (see Tufte, 1993), but simply state that a well conceived graphic will help in the selection of mathematical models and prior densities, while a poor one can easily mislead.

2.3   Data in One-Dimension

When we are examining the observed statistical properties of a single variable, we are typically looking at frequency plots or histograms, either of past data or the current observed data. These provide guidance in the choice of scale and underlying probability distributions that can be used to describe the randomness of the variable in question. They also give a sense of the range and stability of the measurements. If, for example, the data contains many outlying values (outliers), this needs to be considered and may require modification of scales (for example, one-to-one transformations such as the logarithm). If the frequency plot or histogram has more than one mode, this may imply that more than one population is being sampled. This in turn may imply a poor sampling design or the need for a statistical model incorporating a mixture of distributions for more realistic modeling. In general, as long as the basic model under consideration (linear, nonlinear or other models) is supported by the data and the basic hypotheses of interest are not unduly affected, examination of the data and simple modification of the variable scaling is useful. From a Bayesian perspective even more so as the observed data are formally conditioned upon in relation to probability calculations.

   Histogram

   The simplest data graphic examines the frequency of the observed values. The "histogram" shows the relative count for observed values across a set of intervals making up the range of the variable of interest. The shape of the histogram will help support choices and assumptions to be made regarding potential likelihood functions. Histograms drawn from past studies may aid in selecting appropriate prior densities.

2.3.1   Example 1

The following well-known dataset represents a series of calibrated measurements made on a set of eruptions by Old Faithful measured as changes in pressure level per cubic foot. The original dataset can be found in the standard R software package ("faithful" dataset). Histograms of the data using different spaced bins are shown in Figure 2.1. It is obvious that the scale at which we examine data may affect the perceived shape of the frequency distribution.

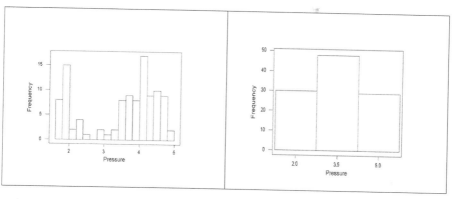

FIGURE 2.1
Histograms with Varying Scales

2.3.2 Example 2

A dataset presented in Manley (2004), p. 2–3, is used here to illustrate basic multivariate data analytic and graphical methods. The initial data is a collection of measurements made in 1898 by H. Bumpus at Brown University on a collection of injured female sparrows after a particularly severe storm, some of whom died. The goal was to examine which physiological variables were related to survival and thus, on some scale, related to the process of natural selection. Six of the variables collected were total length, alar length, length of beak and head, length of humerus, length of keel of sternum and survival status. The data is attached as Dataset 1 and is further examined in the Exercises below. Histograms are shown in Figure 2.2 for several of the continuous variables, both overall and broken down by survival group.

Boxplot

Another plot of interest, especially in the presence of outliers, is the boxplot. This shows a central box containing the middle 50% of the data, along with a line at the median (50th percentile) value. Additional lines stretch out

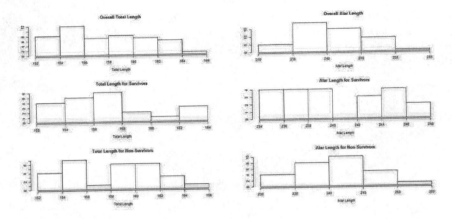

FIGURE 2.2
Variable Histograms

to show minimum and maximum values. See Figure 2.3 where boxplots are shown both overall and broken down by survival group.

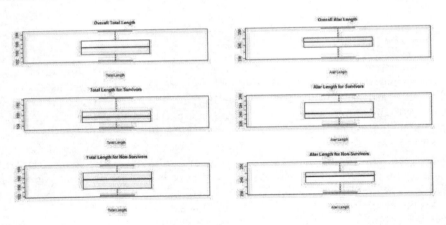

FIGURE 2.3
Variable Boxplots

The ranges, distribution shapes and various statistics in these summaries guide the selection of more formal model components, and the resulting statistics of interest reflect the choice of underlying model and likelihood function. In Part 4 this interaction between data analysis and model selection is examined in various biological and ecological problems.

Transformations

To attain a simpler or more stable representation of the data, it is sometimes necessary to employ data transformations. These typically include the log, square root, inverse or squaring of all values. The scaling of the data is often important, especially in the modeling of biological and ecological variables where re-scaling is often required for the interpretation of data in the context of specific models. The basic underlying rule guiding the application of such transformations is interpretability. All transforms should be $1 - 1$ functions, allowing us to move easily between the new rescaling and the original scale. In Figure 2.4 we show log, square root and inverse transformed histograms of total length and alar length. These transformations may yield a major change in the shape of the distribution or may not. Note that negative or zero values in the data will rule out the use of a log transform. Higher order effects in statistical models (for example interaction) may be removed in some cases by use of a log transformation.

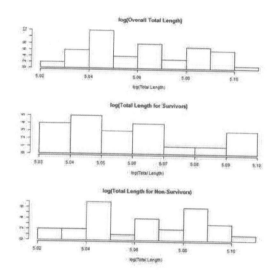

FIGURE 2.4
Log Transformed Variables

We typically prefer a scale yielding a more "normally" shaped density as we understand its properties and limitations. But normality need not be a goal or even a good choice. If the models involve categorical data (for example survival status), a continuous density choice is not useful.

It is worth noting that in larger samples, the importance of the underlying distribution for each variable is somewhat diminished due to the presence of central limit theorems. These have a more obvious effect from the frequentist perspective, but also impact Bayesian methods through the related shape of the likelihood function. Bayesian methods do not formally require large samples, with the relevant posterior densities obtained typically being exact or almost exact in fairly small samples due to advances in computing and related MCMC methods. But obviously, the smaller the sample size, the less justification exists for the assumed shape of the likelihood, the representative nature of the data and perhaps the entire scientific study.

Outliers

There are often anomalies in data, aspects of the data that do not allow a researcher to immediately apply standard models and approaches and fall outside existing expectations. Some typical situations are the presence of heterogeneity (non-constant variation), and outliers (extreme values). Both can arise naturally, but also may signal deficiencies in the sampling scheme employed. This would be the case if the researcher has inadvertently mixed together differing populations or contaminated the sample with extraneous data points.

Mosteller and Tukey (1977) give a listing of potential transformations for a response y, typically $1/y$, $\log(y)$, $\sqrt{y}$ and y^2. The Box-Cox approach uses a data driven approach, assuming a model of the form

$$y_i^\lambda = \beta_0 + \beta_1 x_{1i} + ... + \beta_p x_{pi} + \varepsilon_i$$

and fitting a value to the parameter λ from the observed data. This type of empirical fitting can be helpful, but data transformations have a simple rule; the transformations used must be interpretable in the setting of the scientific problem itself, and this may limit the type of transformation that is practically useful.

2.4 Data Patterns in Higher Dimensions: Correlations and Associations

Continuous Variables

When two or more variables are measured together, perhaps on the same individual subjects or within the same sampling area, we can examine them individually using the data plots mentioned above. But we can also look for evidence of a relationship between them. Most model building is focused on the detection and modeling of relationships between responses of interest (dependent) variables and sets of explanatory variables.

A useful plot is the simple scatterplot, plotting the values of one variable versus another. A multiple scatterplot or matrix plot for several of the variables in Example 2 is given in Figure 2.5. We can see that many of the plots support a linear structure. These plots can also be further broken down by category if desired.

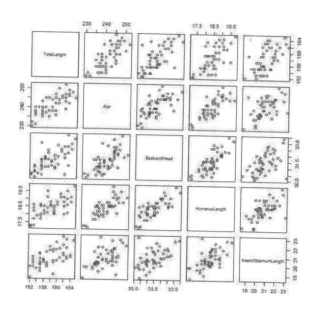

FIGURE 2.5
Multiple Scatterplots

In these plots we typically look for initial evidence of "structure" such as linear or nonlinear patterns. These guide the development of later regression and multivariate models and related likelihoods, which in turn define the setting (ranges, relationships among variables) within which formal prior densities for the model are defined. It is again possible to alter the scale of the various responses, for example using a log transform. The goal is to find a stable, typically linear, relationship that can be modeled.

An often used measure of (linear) association between two sets of measurements x and y is the sample correlation value defined by

$$r = \frac{\sum_{i=1}^{n}(x_i - \overline{x})(y_i - \overline{y})}{\sum_{i=1}^{n}(x_i - \overline{x})^2}$$

where $-1 \le r \le 1$. The closer r is to either 1 or -1, the more aligned the data will be with a straight line (with positive or negative slope respectively). Note that if the data follows a non-linear pattern, for example a parabola, it will give a low r value, even though there is clearly a pattern in the data. Correlations assume a primarily linear pattern in the data to be meaningful. The data should be examined graphically to avoid these types of mis-specification.

If there are more than two variables, a matrix of sample correlation values can be examined. For example with three variables x, y, z the sample correlation matrix is symmetric about the diagonal (since $r_{xy} = r_{yx}$) and looks like

$$R = \begin{bmatrix} r_{xx} & r_{xy} & r_{xz} \\ r_{yx} & r_{yy} & r_{yz} \\ r_{zx} & r_{zy} & r_{zz} \end{bmatrix}$$

where r_{xx} is the sample variance of the variable x. For the set of continuous variables (Total Length, Alar Extent, Length of Beak and Head, Length of Humerus and Length of keel and sternum) in Example 2 we obtain the correlation matrix

$$R = \begin{bmatrix} 1 & .735 & .662 & .645 & .605 \\ - & 1 & .674 & .769 & .763 \\ - & - & 1 & .763 & .526 \\ - & - & - & 1 & .607 \\ - & - & - & - & 1 \end{bmatrix}$$

The observed correlations lie in the range (.607, .735), which along with the matrix plot above, indicates a moderately strong linear structure in the data when the variables are examined in pairs.

As a rule of thumb, most standard approaches to statistical inference and modeling with continuous data are easier to apply and interpret if the underlying structures in the data are linear on some scale. It is therefore important to understand if such simple structures exist in the data or if more complicated models are necessary.

Categorical Data

In datasets where the variables of interest are both categorical, we typically examine a cross-tabulated table of counts and related frequencies. We can also examine the odds ratio and measures of agreement if we are comparing, for example, diagnostic tests. Data that is primarily continuous can always be re-expressed as being above or below a given cut-off value and thus rendered categorical. We examine this briefly in an example.

2.4.1 Example 3

A simulation study is developed to examine the relationship between small animals of two species and the degree to which they are found to contain a given toxic chemical in their blood. The study is conducted for three months in a pre-determined wooded area near a highly used roadway, with comparison to a control group in a more rural area several miles away. The study is in response to environmental concerns that a new gasoline additive (containing

TABLE 2.1
Overall Data

	Chemical	No Chemical	Total
Exposed	30	20	50
Control	20	30	50
Total	50	50	100

TABLE 2.2
Data by Strata

A	Chemical	No Chemical	Total
Exposure	25	5	30
Control	5	25	30
Total	30	30	60

B	Chemical	No Chemical	Total
Exposure	10	10	20
Control	10	10	20
Total	20	20	40

the toxic chemical) will adversely impact small animal species. The data are shown in Table 2.

For this table the odds ratio is $(30)(30)/(20)(20) = 9/4 = 2.25 > 1$, with a 95% confidence interval of $[1.01, 5.01]$, implying no significant association between Exposure level and the presence of the Chemical at this sample size. This may be further modeled using additional explanatory variables and a regression setting, for example logistic regression or Poisson regression (see Part IV).

We can also initially examine the effect of a third categorical variable on this data. If, in the dataset above, we keep track of whether each animal is of sub-species A or B, we will have two tables with comparable descriptive statistics. Here the OR $= 25$ (95% confidence interval $[6.43, 97.2]$) for Table 2.2A and for Table 2.2B, OR $= 1$ (95% confidence interval $[0.29, 3.45]$). Note that for this data, overall there is no relationship, but there is evidence of relationships in one of the sub-species. This is an example of Simpson's paradox (Simpson, 1951), the idea that overall patterns of association may not be reflected within individual (conditional) strata.

2.5   Principal Components Analysis

A more involved approach to rescaling data in a multivariate setting that makes use of correlations betwen variables is principal components. A useful reference is Joliffe (1986) and the approach dates to Pearson (1901). This rescaling is typically available in most software packages. Writing the set of n subject measurements on each of m variables $\mathbf{X}_i$ as columns in a data matrix, we have the n by m matrix $\mathbf{X} = [\mathbf{X}_1, \ldots, \mathbf{X}_m]$. A sample correlation or sample variation matrix $\mathbf{S}$ can be obtained for this matrix. Let the ith diagonal element of $\mathbf{S}$ be given by s_{ii}. The total variation in the data can then be defined as $\mathbf{TOT_X} = \sum_{j=1}^{m} s_{ii}$.

$$\mathbf{S} = \begin{bmatrix} s_1^2 & s_{12} & \cdots & \cdots & & s_{1m} \\ - & s_2^2 & \cdots & \cdots & & s_{2m} \\ - & - & \ddots & \cdots & & \\ - & - & - & s_{m-1}^2 & s_{m-1m} \\ - & - & - & - & s_m^2 \end{bmatrix}$$

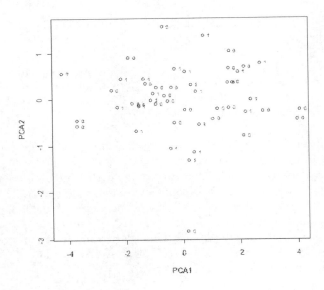

FIGURE 2.6
Plot of First Two Principal Components

To apply the principal components breakdown, an eigenvector based linear

TABLE 2.3
Principal Components

	X_1	X_2	X_3	X_4	X_5	Percent Total Variation	Cumulative Total Variation
Y_1	−.452	−.462	−.451	−.471	−.398	.72	.72
Y_2	−.05	.300	.325	.185	−.876	.11	.83
Y_3	−.690	−.340	.454	.411	.178	.08	.91

transformation of the columns in $\mathbf{X}$ is derived. The resulting columns $\mathbf{Y} = [\mathbf{Y}_1, \ldots, \mathbf{Y}_m]$ are uncorrelated and the resulting m by m sample variation matrix $\mathbf{S_{X^*}}$ for $\mathbf{Y}$ is diagonal with $\mathbf{TOT_Y} = \mathbf{TOT_X}$. The equations defining this new set of variables, the principal components, can be written:

$$\mathbf{Y}_1 = a_{11}\mathbf{X}_1 + a_{12}\mathbf{X}_2 + \ldots + a_{1m}\mathbf{X}_m$$

$$\vdots$$

$$\mathbf{Y}_m = a_{m1}\mathbf{X}_1 + a_{m2}\mathbf{X}_2 + \ldots + a_{mm}\mathbf{X}_m$$

where the coefficients a_{ij} are elements of the ith largest eigenvector of $\mathbf{X} = [\mathbf{X}_1, \ldots, \mathbf{X}_m]$. Typically the subset of principal component variables accounting for at least 80% of $\mathbf{TOT}$ are retained for further interpretation and analysis. As the resulting principal components $\mathbf{Y}_i$, are each linear combinations of the original variables and uncorrelated, they can often be interpreted in a useful manner, but not always. Table 2.3 gives the loadings of the first three PCA variables in relation to the original bird measurements. The first PCA can be interpreted as giving the overall average of the variables, often relating to the overall size of the animal. The second PCA is basically the difference between the weighted average of X_2, X_3, X_4 and X_5 and the third PCA the difference between the weighted averages of X_1, X_2 and X_3, X_4.

Plotting the first two PCA values is often used as a method for detecting outliers in multivariable datasets. In Figure 2.6 this is given with the individual points labelled by survival outcome. There seems to be a pattern of non-survival (0) in the more extreme values of each two principal component. This may imply that these scales are relevant to the process of natural selection. Such data analytic insights can be very useful in the attempt to understand patterns in a given dataset.

References and some exercises relevant to this chapter may be found at the end of Chapter 4.

3

Some Basic Concepts in the Design of Experiments

3.1 Design of Experiments and Data Collection

Good science often requires the detailed planning of experiments, observational studies and data collection procedures. Quality data is the result of planning and careful collection of subjects, clinical measurements, environmental and demographic contextual data, exposures, responses of interest and any other data of interest. A major component of standard frequentist statistics is the theory of design of experiments, a detailed study of how to incorporate randomness and resulting probability into planning a scientific study. The basic goal is to utilize randomization to average out potential bias in the study, reduce overall variation where possible and identify and parse out independent sources of variation among the subjects and variables. Replication of the experiment, as well as follow-up through time are also important considerations in developing experimental and study designs.

Technically these issues are pre-experimental in nature and reflect randomization as a key element in the selection of subjects or assignment of treatments. Many resulting frequentist procedures have the interesting property that the optimality in question does not reflect the actual data collected, but the underlying randomized data collection procedure. Bayesian analysis tilts in the other direction, conditioning completely on the observed data, with probability elements transferred directly to the parameter space via the use of a prior density. However as the practical goal of design of experiments is to achieve representative samples, this is also useful from the Bayesian perspective. The careful elicitation of a prior density is also a key component of the design of experiments in the Bayesian context. A further goal of designed experiments is to lower variation in the observed data. This is typically accomplished, for example in a case-control study, by matching cases and controls by gender, age or other variables. All these are important considerations in generating the data that will be analyzed from both frequentist and Bayesian perspectives.

In some settings there are restrictions on the sampling structure available for selecting the observed sample. In these settings optimal randomization remains a goal, for example using latin square designs or replication at various levels of a design variable (blocking). If there are more complex limitations on the design such as spatial sampling issues or hierarchical structures where sampling is carried out within nested settings (eg. census tracts), then the goal of sampling is to provide a representative sample which reflects randomization and is unbiased with minimal variation.

In terms of actual analysis, estimation and testing, the linear model is applied, $y = X\beta + \varepsilon$. In the context of design of frequentist experiments, ANOVA tables are used to parse out and compare relevant sources of variation and significance regarding values of β. A wide variety of basic designs can be placed in this simple format, subject to various restrictions. These model and design related considerations extend beyond the assumption of normal error via the generalized linear model setting which uses the maximum likelihood approach to estimating parameters and determining optimal fitted models. See Christensen (1987) for a detailed review. This is discussed in Part III and in the applications that make up Part IV.

Sample size issues are also important in the development of study design. In many settings power calculations in relation to expected levels of accuracy in estimation are used to justify sample sizes. Bayesian versions of these exist (Pham-Gia and Turkkan, 1992). Typically a larger sample size will yield smaller standard errors and greater power and accuracy in estimation. A general reference for the determination of sample sizes via power calculations can be found in Cohen (1988). It is worth noting the possibility of over-sampling; large samples can achieve levels of accuracy that generate p-values which reject almost all hypotheses, no matter how trivial, as the standard errors go to zero. In high-dimensional databases this is a concern and many standard methods, frequentist and Bayesian require modification in such settings. This remains an open issue in relation to large database analysis.

As a general guideline, the following technical issues should be carefully reviewed in designing an experiment, whether using Bayesian or frequentist analyses:

- The need for controls and how they are used in the study (as a reference group or matched to cases for example).

- The level of variation present in the data along with the sampling variation of the estimators to be employed.

- The sample size, desired power and accuracy of estimation.

- Issues related to replication and controlling variations such as blocking and within subject follow-up.

- Design related issues such as the presence of hierarchical sampling structures, restrictions on randomization, possible confounding.

- The mathematical model to be used that will incorporate the sources of variation, levels of replication, treatment groups, exposures and other potential factors. This can be linear in basic structure or more complicated. Issues such as stratification to avoid confounding, the presence of higher order effects, nonlinearity and interaction effects need to be carefully considered.

- The elicitation and modeling of prior belief in relation to the model-data combination being employed in the study.

3.1.1 Simpson's Paradox

The effects of secondary variables in defining subgroups within the overall dataset is an important consideration when planning the collection of data in a study. Surprisingly overall patterns in the data can be nullified or even reversed when the data is broken down into subgroups for analysis. These patterns are typically referred to under the general title of Simpson's paradox or aggregation effects or association reversal. See for example Brimacombe (2014).

These effects occur when factors affecting responses in question are not well understood. In such a situation key variables may not be collected as part of the study design and left out of the statistical model and subsequent analysis. Therefore the data is not properly subdivided into important subgroups. Simpson's paradox arises here when marginal or overall response patterns in the aggregated data are typically null or the opposite of conditional within group response patterns. This is not really a formal logical paradox. It is more a design flaw or limitation in the science underlying the study itself.

Mathematically, Simpson's paradox can be stated;

$$Y \sim X$$
$$Y \nsim X, W = w_1$$
$$Y \nsim X, W = w_2$$

where $\sim$ denotes association. This states that the overall association observed between the response variable (Y) and an explanatory variable (X) does not hold within the strata defined by a third variable (W). These considerations are also relevant to issues regarding causation and conditional independence generally. They occur for continuous and discrete random variables. For more background, see for example Julious and Mullee (1994).

3.1.1.1 Example 1

Consider a study examining a linear relationship between a set of observed gene expression values y_i and dosage levels of a specific drug x_i. A secondary variable w_i reflecting body metabolism is suspected of being associated with this relationship. A simple model, $y_i = \beta_0 + \beta_1 x_i + \varepsilon_i$, is fit to all the data and also within strata of the w_i variable. The data and fitted model results are shown in Table 3.1. The regression coefficient for the slope is significantly positive in the overall data but is significantly negative in the subgroups.

TABLE 3.1

Simpson's Paradox in Regression Data

Group 1														
y	15	12	14	13	10	5	7	6	5	4.5	5.8	5	6	5.6	4.3
x	3	7	6	5	4	5	6	9	10	11	9	8	8.9	9	9.3

Group 2														
y	32	30	31	28	26	25	28.9	27.2	25	23	22	20	19	20.1	17
x	20	24	25	27	28	29	32	33	32	30	30	32	33	34	35

	Overall	Group 1	Group 2
$\widehat{\beta}_1\ (\pm s_{\widehat{\beta}_1})$	0.63($\pm$0.098)	$-1.17(\pm0.29)$	$-0.88(\pm0.19)$

Obviously aggregated data can give misleading results where subgroups or stratified data are of interest.

3.1.1.2   Example 2

Categorical summaries are as susceptible to this effect as well. Assume here that the association between a Phenotype (yes/no) and a Gene Expression level above a given threshold (yes/no) is examined in the presence or absence of a specific epigenetic exposure. The overall contingency table and tables for each respective exposure group are shown in Table 3.2.

If the bioassay effect is not incorporated we see that the estimated overall effect is misleading with the odds ratio altering from greater than 1 to less than 1.

3.1.1.3   Path Analysis

Path analysis examines collections of potentially correlated variables in terms of their relation to the response of interest. A path analysis model is typically expressed in terms of variables which are both measured and not directly observed (latent). This usually reflects a theoretical model. See Wright (1921).

For example, consider a setting where two background environmental variables are related to two gene expression variables and two resulting phenotypes. The path diagram for this setting is graphically expressed in Figure 3.1.

The correlations between each pair of variables of interest is defined along the path connecting them. Unknown correlations can be given values to examine the robustness of the overall correlations obtained. Correlations are obtained between any two elements in the path diagram by multiplying the correlations of all the elements along each connecting path. If there are multiple paths connecting the variables of interest, correlations are summed across all individual paths to obtain the overall correlation.

Here the correlation between Epigenetic factor 1 and Phenotype 1 is given

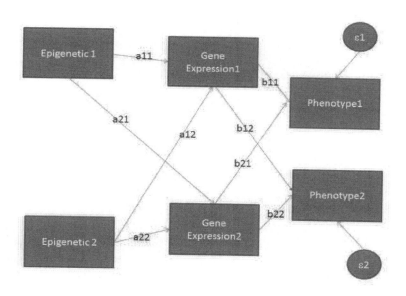

FIGURE 3.1
Path Analysis Diagram

TABLE 3.2

Simpson's Paradox with Categorical Data

Overall				
		Phenotype		
Gene Expression	Yes	No	Total	Percent Yes
Yes	30	30	60	50
Nol	15	24	39	38
Total	45	54	99	$(OR = 1.6)$

Exposure Yes				
		Phenotype		
Gene Expression	Yes	No	Total	Percent Yes
Yes	28	22	50	56
No	7	3	10	70
Total	35	25	60	$(OR = 0.54)$

Exposure No				
		Phenotype		
Gene Expression	Yes	No	Total	Percent Yes
Yes	2	8	10	20
No	8	21	29	28
Total	10	29	39	$(OR = 0.65)$

by $a_{11} \cdot b_{11} + a_{21} \cdot b_{21}$ and the correlation between Epigenetic factor 2 and Phenotype 1 is given by $a_{22} \cdot b_{21} + a_{12} \cdot b_{11}$.

Path analysis allows researchers to visualize potential relationships among a set of variables. Note that Simpsons paradox can occur here when the analysis is applied to various subgroups and the identified path networks differ in structure from the overall path analysis.

3.1.2 Replication in the Design and Analysis of Experiments

Replication as a component of experimental design allows for better modeling of random variation in the response of interest and helps clarify the significance of results. This is especially true in newer pilot areas of research, where understanding random variation is often a challenge due to smaller sample sizes and basic scientific issues of scaling and calibration. Without the ability to carefully model random variation, it is difficult to develop a predictive model that has good accuracy and stability.

The stability of results in replicated settings is fundamentally an experimental issue and it is only once stability and basic calibration are attained that statistical issues such as Type I error or multiple comparison corrections should be considered. High quality for the observed sample is often tacitly assumed, and this may not be the case.

Statistical methods that condition completely on the observed sample inherit the difficulties of the particular sample observed, for example bias or heterogeneity and require increased use of replicated experimental data to validate inferences. These include Bayesian and resampling based approaches. Modified definitions of significance in these settings is not the issue. Experimental replication is the true test of a scientific result.

Technically, replication of the experiment allows for greater understanding of the standard error both overall and in the assessment of mean differences between treatment groups. Replication can also be employed to generate goodness-of-fit tests for aspects of the model itself, with pure error estimates from planned replications playing a key role. While having controls present in an experiment is often viewed as necessary to claim good design, replication should also be present.

Random variation in experimental studies can occur for example at the levels of individual bioassay, shared platform, researcher, sample processing, clinic, all of which affect the stability of the context within which individual responses are observed. These factors can affect the modeling of variation in relation to both control and treatment subjects, leading to the confounding or biasing of subsequent statistical analysis. Modifying a significance level cutoff in a single observed trial will not help matters when the assumed experimental model is not in sync with the stochastic process underlying the collection of the data.

The design of a successful experiment mirrors the underlying science and contains experimental structures that allow for the clear testing of hypotheses subject to reasonable model assumptions and replicability.

3.1.2.1 Replication and Modeling

In experimental settings statistical models employ variation to assess whether, after accounting for random source(s) of error, there is a difference between subject groups, for example cases and controls. This is often a challenge as underlying variation may have several causes. In genetic studies for example these may include naturally occurring variation at the genetic and cellular level, bioassay variability, as well as translation and alignment effects. The formal modeling of these sources of randomness and their decomposition in relation to various experimental factors is essential in interpreting the experiment in question. ANOVA is often useful in parsing out these sources of variation.

While replication within subject is sometimes required to model follow-up or growth patterns over time, replication of the entire experiment is of interest here. Let's look at some effects of replication in regard to basic statistical approach and the interpretation of results.

TABLE 3.3
Replicated Study Data

	Bioassay 1	Bioassay 2
Controls	1.5,2.3,1.7,1.2,2.1,2.2,2.3,1.6,1.7,1.8	2,2.4,2.2,2,2.1,2.1,2.2,2.4,2.2,2.3
Treatment 1	3.4,4.2,5.2,4.9,6.1,4.3,4.2,3.9,5.3,5.4	2.4,2,2,2.2,2,2.1,2,2.3,1.9,2.1
Treatment 2	6.1,7.2,8.1,7.9,9,8.5,8.3,9.4,8.9,8.8	2.1,2,2.7,2.1,2.2,2.1,2.6,2.3,2.1,2.6

Example

Consider a simulated study with three groups; one control and two dosage levels (Treatments 1 and 2) of a new treatment to enhance the average growth of red blood cells. Two trials are conducted. The response of interest is growth after a given period of time on a standardized scale (0–10), and a distinct bioassay platform is used for each trial. Subjects are genetically identical mice given the same basic diet. This design allows for smaller sample sizes and often low variation. The basic dataset here is given in Table 3.3.

3.1.2.2 Fully Replicated Design in Single Overall Model

The ANOVA model serves here as the basis for decomposing variation when the entire experiment is completely replicated as a component of the overall planned experimental design. In most studies of this type replication is a formal aspect of the study and the experiment is replicated across all levels of a so-called blocking variable. This design tends to go under the name of (completely) randomized blocking design. Challenges arise when the blocking is not, in some sense complete, due to resource limitations or other restrictions. Such non-completely crossed designs include the split plot design, latin squares design and other variations of restricted randomization. See, for example, Hicks (1982).

The data above can be viewed in such a blocking format with the blocking variable being the two different bioassay platforms "Bioassay 1" and "Bioassay 2" respectively. The ANOVA format uses the decomposition of random variation into sum of squares components (sum of squares regression (SSR), sum of squares error (SSE) and total sum of squares (SST)) to relate the response to the explanatory factors in question. It is particularly useful in the context of replicated observations.

The model for a 1-way ANOVA with additional blocking variable (assume completely randomized) can be written:

$$y_{ijk} = \mu + \gamma_i + block_j + \varepsilon_{ijk}$$

with ε_{ijk} i.i.d. $N(0, \sigma^2)$, μ a baseline average level and γ_i the treatment effects. This can also be interpreted formally as a two-way ANOVA without interaction, with an F-test derived for overall main effect and blocking variable. This is often a useful approach as it provides an overall sum of squares value based on the overall sample and thus typically provides a more accurate estimate

TABLE 3.4

ANOVA for Overall Data with Blocking

Source	df	SS	MS	F	p-value
Treatment	2	105.40	52.70	25.46	0.0001
Blocking	1	111.52	111.52	53.87	0.0001
Error	56	115.92	2.07		
Total	59	332.85			

for σ^2. This implies increased statistical power. However, this assumes that the bioassay subgroups are comparable in terms of design and experimental context.

For our example, we obtain the ANOVA table and the treatment effect is significant. See Table 3.4.

Replication of Complete Experiment in Distinct Models

In settings where pilot data is to be collected, calibrated and assessed, a more conservative approach to replication is sometimes more useful. Here the experiments in question may not be well understood and difficult to view as truly comparable. We now analyze the results from the different bioassays as separate, independent experiments. This is more conservative as potentially significant differences in each analysis do not benefit from the larger overall sample size and smaller SSE available if viewed as a single experiment with blocking.

In research settings where a high percent of results do not replicate, such a conservative approach is to be seriously considered before a finding is reported. Attempting to maximize statistical power is not the primary goal when the underlying experiment requires further calibration. Once results are consistently observed across bioassay platforms and other secondary sources of variation, they can be made the basis of further detailed study.

Assume a 1-way ANOVA model $y_{ij} = \mu + \gamma_i + \varepsilon_{ij}$, with $i = 1, ..., 2$ and $j = 1, ..., 10$ with the errors for each observation assumed independent, $\varepsilon_{ij} \sim N(0, \sigma^2)$. For the example here we can run two separate ANOVA based analyses for each Bioassay subgroup respectively (Table 3.5). The results for both trials should agree for the overall inference to be viewed as consistent and stable. Here they do not. Indeed half the collected data (Bioassay 1) implies a result while the other half (Bioassay 2) does not.

Comparing these results to the first analysis above we see that combining results in the presence of blocking may provide a false sense of comfort as greater sample size and resulting higher power and lower standard error mask the underlying experimental instability. When results are not well understood, separate analyses can be more informative. Note that if additional replications of the experiment continued to give a similar pattern of disagreement, there would be no additional evidence of a result, but the SSE related to the over-

TABLE 3.5

ANOVA for Each Replication Group Separately

Block 1					
Source	df	SS	MS	F	p-value
Treatment	2	204.29	102.14	173.79	0.0001
Error	27	15.87	0.59		
Total	29	220.16			
Block 2					
Source	df	SS	MS	F	p-value
Treatment	2	0.16	0.081	2.18	0.13
Error	27	1.01	0.037		
Total	29	1.17			

all blocking related analysis would diminish further and the original overall misleading main effect p-value become even more significant.

3.1.2.3 Significance Issues

Modern methods available for the determination of significance can themselves be highly conditional on the observed data and as such will reflect the biases, errors and design defects present in the observed dataset. Both Bayesian and computer-based frequentist bootstrap assessment methods (also randomization and permutation tests) are conditional on the observed data. Their growing use argues for increased replication as a means of validating significance measures such as p-values and posterior odds ratios.

The frequentist approach to the assessment of evidence often reflects properties that assume averaging over potential experiments not yet conducted. This can lead to very conservative significance levels. Note that the 0.05 cutoff for significance is often attributed to R.A. Fisher, but his view was flexible, seeing the 0.05 cutoff or any significance cutoff as being related to the context of the science in question and best left to the researcher's judgement. See for example Fisher (1955). Indeed at times he was quite insistent that dogmatic interpretation and reporting of significance cutoffs and statistical results in general was not appropriate. This has often been misrepresented.

One of the challenges in new areas of research such as cancer genomics is the limited existing scientific understanding of the mechanisms triggering and sustaining genetic expression, subsequent proteomic and cell development and the fact that there may be multiple pathways. In such settings it may be difficult to link statistical significance with clinical significance. Experimental replication seems more useful to determine clinically relevant results than, for example, arguing for less stringent significance cutoffs for individual trials.

Bayesian methods provide an approach to statistical evidence or significance that condition completely on the observed data. These methods require the asking of slightly different questions of the data, namely, how much did we learn from the observed trial or experiment and how has this updated our beliefs? This differs from more conservative frequentist or experimentalist approaches which ask simply, did the treatment work in this trial or, more precisely, can we show that the null hypothesis of the treatment not working can be rejected. These are different perspectives, both using the likelihood function.

Technically, the Bayesian approach modulates the likelihood function into a probability distribution via the incorporation of formal modeling of prior belief using a prior density for the unknown parameter θ of the model. The probability here is directly on the parameter space. Formally this can be written on a log scale:

$$\ln p(\theta|x) = \ln c + \ln p(\theta) + \ln L(\theta|x)$$
$$L(\theta|x) = \prod_{i=1}^{n} f(x_i|\theta)$$

where $p(\theta)$ is the prior density for θ, typically a population characteristic, $L(\theta|x)$ the observed likelihood with x the observed data and c the constant of integration. From a learning perspective, we have initial prior belief for θ and update this using the likelihood function, giving updated belief in the form of the posterior density $p(\theta|x)$.

The use of the observed likelihood function gives the Bayesian approach a strong theoretical basis in regard to the collection of optimal levels of information from an experiment. However the dependence of the Bayesian result on the observed data through $L(\theta|x)$ also implies that bias or calibration issues in this data directly affect the resulting probability calculations and determination of significance. This approach does not require an averaging argument over potential values for x on the sample space.

Bootstrap frequentist inference (Efron and Tibshirani, 1994) is based on resampling the observed data. In a sense it extends randomization and permutation based signficance procedures to a much broader set of statistical contexts and models. In its simplest form the bootstrap method resamples with replacement the observed sample data y, generating multiple values of the statistic of interest $s(y)$ for each independent resample $y_1, y_2, ..., y_B$. The related histogram for these $s(y)$ values, $F^*(s)$ gives the bootstrap distribution which can be used for estimation and testing as long as the resampling used reflects the null hypotheses of interest. This approach is data dependent and is affected by design issues relating to the collection of the original sample y.

In replicated settings there is often discussion of how to adjust the frequentist significance level to account for multiple comparison related issues. See for example Gaudart et al., (2014). Note that significance based on probability calculations is not the only inferential scale available. LOD scores in genetics for example, based on likelihood functions and posterior probabilities typically are referred to a score of 2.0–3.0 or higher to determine significance. Information theoretic cutoffs based on the Akaike Information Criteria (AIC) can be interpreted on a percent change scale. Results may also be represented in terms of the number of standard errors between expected and observed test statistic values. Indeed in very large high dimensional sample settings where both standard errors and tail area probabilities collapse towards zero for any minor clinical effect size, the appropriate scale for determining significance is an open issue. Note that even when using such alternative assessments of significance, design issues and the need for consistent experimental replication remain essential.

3.1.2.4 Pseudo-Replication in Observational Studies

In settings where a large cross-sectional sample is collected from an observational study or large clinical database, the validation of the fitted model using replication may not be directly possible. A pseudo-replication based approach can be employed using a set of random sub-samples of the overall dataset. Observed cross-sectional data can be prone to bias, confounding and other flaws that can affect statistical analysis.

Most simply, the data are randomly divided in half; one piece becoming the training sample, the other the validation sample. A linear model is then fit to the training sample and the fitted model is then run on the validation sample. If there is a high level of agreement between the observed validation sample values and those predicted by the training sample model, the model is viewed as validated.

There are many ways to create or draw training and validation samples and this is typically carried out numerous times, with random selection being used on a subset of key variables. We can split the observed response (y) and explanatory variable data matrix (X):

$$y = y_1 \mid y_2$$
$$X = X_1 \mid X_2$$

where (y_1, X_1) is the training sample. Then fit an initial model:

$$\widehat{y_1} = X_1\widehat{\beta}_1 = \widehat{\beta}_{01} + x_{11}\widehat{\beta}_{11} + x_{12}\widehat{\beta}_{21} + \cdots + x_{1p}\widehat{\beta}_{p1}$$
$$\widehat{\beta}_1 = (X_1'X_1)^{-1}X_1'y_1$$

and run the same fitted equation on the validation dataset, obtaining predicted values:

$$\widehat{y_2} = X_2\widehat{\beta}_1 = \widehat{\beta}_0 + x_{21}\widehat{\beta}_1 + x_{22}\widehat{\beta}_2 + \cdots + x_{2p}\widehat{\beta}_p$$

which can be directly compared to the observed response y_2 in the validation dataset. Simple measures of agreement between $\widehat{y_1}$ and $\widehat{y_2}$ can be employed, for example correlation or statistics such as $PRESS, BIC, AIC$ or C_p statistics (see Draper and Smith, 1981). As an alternative, the best fitting model for each subgroup can also be obtained and compared both to each other.

This approach can be repeated across a set of random choices of (y_1, X_1) and (y_2, X_2). If the fit is acceptable across these random "replicates" of the data, the basic fitted model can be viewed as stable. This approach can also be applied to time dependent datasets. While not true replication, it provides some measure of stability in regard to the fitted model.

3.1.2.5 Replication and Meta-Analysis

The goal of combining the results of many similar studies motivates the set of methods termed meta-analysis. In some settings, for example epidemiology, there is often a decision to make as to whether researchers should carry out one large real-world study to understand a treatment effect or a series of carefully controlled smaller studies, then combining them into a meta-analysis.

Meta-analysis is a type of replication study, bringing together results from a set of published results and analyzing them as an integrated set of results. It is difficult to do well, and requires combining replicated trials conducted in different experimental or observational settings.

A standard meta-analytic summary of the overall average effect size drawn from m individual studies is often expressed in the form of a pooled or weighted average:

$$\overline{Y} = \frac{\sum_{i=1}^{m} w_i y_i}{\sum_{i=1}^{m} w_i}$$

where y_i is the effect size estimated in the ith study and w_i is the inverse of the variance for the ith study and $Var(\overline{Y}) = 1/\sum_{i=1}^{m} w_i$.

If the collected studies are viewed as a single sample of a larger population of studies, they may be given a random effects interpretation and a formal meta analytic model can be written:

$$\begin{aligned} Y_i &= \mu_i + \sigma_i \varepsilon_i \\ \mu_i &\sim N(\mu, \tau^2) \end{aligned}$$

where the ε_i are *i.i.d.* $N(0, 1)$ random errors and $i = 1, ..., m$. The average effect size here is seen as estimating an underlying overall population effect size μ. This approach gives conservative overall effect size estmates. Formal tests of heterogeneity can be conducted to assess the degree to which the studies are similar in terms of statistical variation. Collections of confidence intervals, called forest plots, can be obtained for mean parameters. Bayesian methods can also be applied if a joint prior density for the $(\mu_i, \mu, \sigma, \tau^2)$ parameter vector is assumed.

Note that the issue of unpublished negative results affects the veracity of all meta-analyses. The so-called file drawer effect or non-publication of negative results, potentially biases many meta-analytic studies. This reflects the often cited bias of professional research journals to mostly publish positive results and often disregard negative results. In some areas this implies that any meta-analytic study based on published data may be biased and overstate the treatment effects in question. See for example Sharpe (1997).

Summary

- The design of experiments affects the resulting interpretation of experimental or observational results and in the early stages this should include careful consideration of study replication.

- Without replication a scientific result should be viewed skeptically. Statistical significance is not necessarily clinical significance and this is especially true in the pilot stages of new areas of research.

- In settings where basic results are inconsistent or unstable, poor quality cannot be overcome by single trial statistical corrections based on methods which are themselves conditional on quality related aspects of the observed data, including their potentially biased nature. Statistical power issues are secondary in these settings. Simply placing replicated results into an overall blocking design setting may not be trustworthy.

3.1.3   Incorporating Expectations and Beliefs

Bayesian methods allow for modification and extension of standard statistical methodology. But there is a price to pay. They require the specification of a prior density describing expectations regarding the possible values of the set of parameters in the model. This is a required component of Bayesian design of experiments. While we may have existing subjective beliefs due to theory, past data, if available, is also a place to begin to examine such beliefs. We also typically use these sources to support the choice of probability model and the likelihood function we intend to use to describe the process generating the study data.

The key inferential and information processing tool the posterior density, was defined earlier and can be written:

$$p(\theta \,|\, data) = k \cdot p(\theta) \cdot L(\theta \,|\, data)$$

where $L(\theta \,|\, data) = c \cdot \Pi_{i=1}^{n} p(y_i; \theta)$ is the relevant likelihood function, $p(\theta)$ is the prior and k the constant of integration. If placed on a log scale, we have:

$$\ln p(\theta \,|\, data) = \ln k + \ln p(\theta) + \ln L(\theta \,|\, data)$$

and

$$\frac{d}{d\theta} \ln p(\theta \,|\, data) = \frac{d}{d\theta} \ln p(\theta) + \frac{d}{d\theta} \ln L(\theta \,|\, data)$$

and we can see that the log-prior density serves as a baseline regarding our knowledge regarding θ, we then alter this by adding the log-likelihood which reflects the model-data information from the observed sample and obtain our updated beliefs, the posterior density. If we further examine the first order derivative as a measure of the rate of change in the posterior and thus a type of learning measure, we see that the rate of change in the log posterior is a function of both the rate of change in the log prior and the frequentist Score function

$$S(\theta) = \frac{d}{d\theta} \ln L(\theta \,|\, data).$$

The properties of the likelihood function underlie much of the learning related aspects of posterior based inference. Indeed if the second order derivative is examined we see that when the prior is relatively flat the log convexity of the posterior depends on the log convexity of the likelihood function, a result which also drives convergence to Normality. A well-behaved posterior density reflects a well-behaved likelihood function.

In the areas of ecology and biology there are both in-depth empirical studies and theoretical models available to guide prior knowledge and expectations. These provide the contexts and initial models within which statistical inference is carried out. This is very useful from the perspective of Bayesian statistical application, as the basic variable set, regression structures, ranges for variables, distributions, parameterizations are often available to researchers and can guide the selection of likelihoods and prior densities. While it is always good scientific practice to be as objective as possible towards the analysis of data and the modeling and testing of population characteristics (parameters), good analyses require informed contexts and assumed models within which to interpret data. In this book we emphasize a practical middle ground between assumptions and empirical justification.

References and exercises relevant to this chapter may be found at the end of Chapter 4.

4

Prior Beliefs and Basic Statistical Models

4.1 Selecting Prior Densities

The practical expression of existing belief in the context of Bayesian statistical modeling is in the assumed shape of the prior distribution. There are several specific approaches to selecting priors. These range from very subjective, almost free form approaches, to empirical methods based on previous datasets, properties of the likelihood function, and properties of overall model fit. More extensive discussion than that offered here can be found in Bernardo and Smith (1994).

4.1.1 Subjective Priors

Strictly speaking, in general Bayesian theory, any proper (and sometimes improper) density is acceptable as a prior density if it truly reflects the beliefs of the researcher. All that is required is a properly behaved resulting posterior distribution. The selection of a prior density is an attempt to formalize the beliefs of researchers in regard to the potential or expected values of the parameters in question.

Here we briefly mention the concepts of exchangeability and utility. The concept of exchangeability is a useful concept underlying the formal development of priors and how information is processed. When there is more than one parameter, say θ_i, $i = 1, ..., p$ about which to consider and develop prior beliefs, the order of the parameters should typically not matter when learning about them and modelling associated prior beliefs. Finite exchangeability is taken to formally assume

$$p(\theta_1, ..., \theta_p) = p(\theta_{(1)}, ..., \theta_{(p)})$$

for any re-ordering $\theta_{(1)}, ..., \theta_{(p)}$. This property is viewed as desirable when modelling from the Bayesian perspective and models can be examined for this property. It is not as strong an assumption as independence.

The concept of utility has a long history in economic thinking and the modeling of how individuals behave and develop preferences and associated probabilities for these preferences. It is a key concept in the formal modeling of human behavior when there are several potential outcomes and a choice is to be made among them. Subjective beliefs can be assessed in these terms. As mentioned above, key work linking the concept of utility and prior density functions, as an approach to the expressing of beliefs in a coherent manner, was given by de Finetti (1931) and is a theoretical justification for the use of prior probability densities as adequate models of prior belief and preferences (Bernardo and Smith,1994, Chapter 2).

4.1.2 Previous Likelihoods: A Source of Prior Information

Scientific investigation is rarely conducted without a well-defined context. Often there is a long history of research and the slow accumulation of results. The Bayesian approach, which requires information on the relevant likelihood to be employed, as well as a prior density, depends heavily on pre-existing studies and defined likelihoods from previously fitted models. While we can, strictly speaking, claim ignorance and utilize non-informative priors, when we have previously obtained likelihood functions, we have a very good indication of the type of density the data supports for the parameter in question. In most areas of biology and ecology, parametric theoretical models exist, as do many empirical studies to support such a context for the application of Bayesian methods.

If there exist several previous studies, a smoothed histogram of previous maximum likelihood estimates may be useful to provide an overall empirical summary of existing knowledge regarding the parameter of interest, and thus a prior density for the current posterior density. Boos and Monahan (1986) examine these types of approaches. In larger samples, due to the central limit theorem, we would expect a normal shape for the density, on some scale. In smaller samples this may not be the case, but both re-scaling of the response and re-parameterizations (re-scaling of the population characteristics themselves in the model) may help.

4.1.3 Jeffreys Prior

Another likelihood based approach to deriving prior densities, is to use the properties of the observed likelihood function. Harold Jeffreys, a geologist by training, defined a prior density related to the Fisher information, which is itself based on the observed likelihood function. The Fisher information reflects the average curvature of the log-likelihood about the maximum likelihood estimate and can be written:

$$I(\theta) = E\left[-\frac{\partial^2 \ln L(\theta; y)}{\partial \theta^2}\right]_{\theta=\widehat{\theta}}$$

Large sample arguments in regard to the *m.l.e.* $\hat{\theta}$, where the likelihood satisfies regularity conditions, give the result:

$$\hat{\theta} \sim N\left(\theta, \frac{1}{I(\hat{\theta})}\right)$$

As defined by Jeffreys a useful, locally non-informative prior is obtained by setting the prior equal to the inverse of the asymptotic standard error. Thus the effect of this prior density selection is having a relatively flat, somewhat invariant, prior about the mode of the likelihood where the likelihood is pronounced and we can expect the likelihood function to dominate the posterior density. Formally the Jeffreys prior is given by:

$$p(\theta) = \left|I(\hat{\theta})\right|^{1/2}$$

This is more of an empirical Bayes approach as it is based on the observed likelihood and therefore reflects both the assumed model and the observed data. As it reflects the Fisher information and related Cramer-Rao information bound (the asymptotic variance), it is not really reflecting pre-existing subjective belief. It also may not be directly based on previously existing studies, unless these have informed the choice of the likelihood.

Note that in larger samples we may also leave out the idea of taking expectations and replace $I(\hat{\theta})$ with $J(\hat{\theta})$ the observed Fisher information. A formal displacement argument can also be made in relation to further justifying the Jeffreys prior (Box and Tiao, 1973, p. 36).

When θ is not a scalar, but rather a p-dimensional vector parameter $(\theta_1, \theta_2, ..., \theta_p)$, the Jeffreys prior still applies, but covariance terms between the parameters may need to be determined if prior information is not available independently for each individual parameter θ_i. The Jeffreys prior is often not the theoretically optimal choice of prior and there is a long literature examining such automatic choices of relatively flat priors in Bayesian methods. The Jeffreys prior however is interpretable and useful in a wide variety of likelihood based models.

For the case of a Normal model with unknown mean θ and standard deviation σ, the likelihood and related Jeffreys prior are given by:

$$L(\theta, \sigma^2 | y) = c \cdot (1/2\pi\sigma^2)^{n/2} \exp(-(1/2\sigma^2) \sum_i (y_i - \theta)^2)$$

$$p(\theta, \sigma^2) \propto \frac{1}{\sigma^3}$$

Here the prior is improper (it does not integrate to one).

For the case of the Binomial distribution, with y as the number of successes out of n independent trials, each with probability θ of success, we have the likelihood, Jeffreys prior and resulting Gamma posterior density (plotted) given by:

$$L(\theta|y) = c \cdot \theta^y (1 - \theta)^{n-y}$$

$$p(\theta) = Beta(.5, .5) = \frac{\Gamma(1)}{\Gamma(1/2)\Gamma(1/2)} \theta^{-1/2}(1 - \theta)^{-1/2}, 0 < \theta < 1.$$

In such settings the Fisher Information is often a good start to measuring overall variation in the model and in this regard, the Jeffreys prior, which is designed to be minimal when the Fisher information is greatest, is often a good first approximation to priors with system-wide properties. But it is important to remember that in many ways this approach to Bayesian prior selection is more about modulating the likelihood distribution than it is about prior belief and learning behavior.

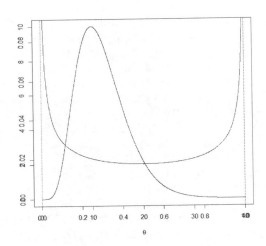

FIGURE 4.1
Gamma Posterior Density with $y = 5$ and Gamma Prior with $\alpha = 1$ and $\beta = 1$

4.1.4   Non-Informative and Improper Priors

In some settings, usually where little is known regarding the population characteristic under study, the use of a flat, non-informative prior density is a typical first step, $p(\theta) = c$. If however, the parameter under consideration is unbounded, then the flat prior will not integrate to one and is not a proper density. This has not stopped researchers from employing such densities, and this is similar to taking the likelihood function itself as the posterior density. However, the use of improper priors can result in posterior densities that themselves do not integrate properly. As well, the order of integration, when deriving marginal posteriors, which technically should not matter, occasionally does matter when using such priors (Stone and Dawid, 1972). Caution should be exercised in relation to such prior selection.

4.1.5   Conjugate Priors

Conjugate priors are prior densities that, for a given likelihood, give a posterior function that is similar in form to the prior. In practical settings, it is rare that we have such strong confidence in our choice of prior densities. We tend to have stronger information regarding the choice of likelihood, though in some time series related settings the likelihood itself may not be completely defined. Most applications of conjugate priors are in the setting of exponential family based likelihoods (Normal, Poisson, Gamma, etc. see below). While it is sometimes seen as an artifact of the days before advances in multiple integration via application of Markov Chain Monte Carlo (MCMC) methods, the interpretability of conjugate priors can be useful. They often result in posterior densities parameterized in a manner that allows for interpretation of respective prior and data contributions to the posterior density.

Consider an example where we are modelling counts (cells, number of animals in a given species, number of subjects that recover from a disease etc.) in a specific time period. This is a setting where the Poisson probability distribution can be used to build the likelihood function. Letting y be the number of observed events in the time period, the likelihood is given by:

$$L(\theta; y) = c \cdot \frac{e^{-\theta}\theta^y}{y!}, \theta > 0, y = 0, 1, 2, 3, ...$$

The related prior density will be over the region $(0, \infty)$. A useful and flexible choice in this situation is typically the Gamma density. This is given by:

$$p(\theta) = \frac{\theta^{\alpha-1}e^{\theta/\beta}}{\Gamma(\alpha)\beta^\alpha}, \theta > 0, \alpha > 0, \beta > 0 \tag{4.1}$$

with α and β assumed to be known, and $\Gamma(\alpha)$ is the gamma function. We write this density as $Gamma(\alpha, \beta)$ where α and β are chosen hyper-parameters and typically reflect subjective belief. This prior has mean $\alpha\beta$ and variance $\alpha\beta^2$. The shape of this distribution depends on α only. With $\alpha \leq 1$, it has a single-tailed shape, with $\alpha > 1$, a two-tailed shape. The posterior density for this prior and likelihood combination is given by:

$$p(\theta \,|y) = c \cdot \theta^{y+\alpha-1}e^{-\theta(1+1/\beta)}$$

which is proportional to the Gamma prior density with adjusted parameters $Gamma(y + \alpha, (1 + 1/\beta)^{-1})$. This implies that $p(\theta)$ is therefore the conjugate prior for the Poisson likelihood. This similarity in form defines the conjugate prior idea. The magnitude of α and β represent the effects of the prior density on the posterior from which inferences are drawn. Central 95% regions can be reported to estimate θ.

4.1.6 Reference Priors

A general learning model approach to Bayesian statistics can be employed. Here the improvement in the information available to the researcher when going from prior to posterior distribution is the focus. The most developed of these approaches is the reference prior approach. The reference prior, under specific conditions, maximizes the amount of missing information between the posterior and likelihood, on a specific functional scale, averaging over elements of the sample space. While well defined in single parameter problems, the reference prior is only well defined in multi-parameter problems relative to an ordered parameterization, see Bernardo and Smith (1994) Chapter 5 for an overview. The determination of the reference prior in general settings is challenging and we do not apply the approach here.

4.1.7 Elicitation

There is a large literature on the formal ellicitation of prior beliefs and much of this is reviewed in Bernardo and Smith (1994) and Garthwaite et al. (2005). The approach is typically based on underlying notions of relative utility and is often similar to describing a person's utility function, but in regards to potential values for the population characteristic parameter of interest. Preferences are viewed as based in utility, but expressed on a related probability scale more appropriate for statistical analysis. In ecological and biological settings there is typically a strong theoretical aspect and the approaches to selecting priors given above are typically acceptable. Elicitation is difficult to do well, but may be a useful approach to defining priors based on carefully assessed expert opinion. Elicitation is not developed in detail here.

4.2 Model Nonlinearity and Prior Selection

In many ecological and biological settings the underlying theoretical relationship to be placed in a statistical setting is nonlinear. In such settings, the likelihood function can be complicated and the respective selection of prior densities challenging. The likelihood function can be highly non-normal in shape and require re-parameterisation to be useful in global maximization.

If the nonlinearity is rooted in a time related model, then the instabilities of the underlying mathematical model present challenges to simply attaining a stable mode for the likelihood or posterior. In such settings the choice of prior is often undertaken not to reflect personal beliefs, but to attain a stable

TABLE 4.1

BOD Data

Oxygen Demand (y_i)	8.3	10.3	19.0	16.0	15.6	19.8
Time (x_i)	1	2	3	4	5	7

posterior density which can then be used to study the underlying model-data combination.

4.2.1 Example 4: BOD Example

An asymptotic growth model is applied to the BOD dataset found in Bates and Watts (1988) and is given by:

$$y_i = \beta_1(1 - \exp(-\beta_2 x_i)) + \varepsilon_i.$$

This model relates oxygen demand to time through a nonlinear regression model. The data are given in Table 4.1 and a plot of the data (shown earlier) is reproduced here.

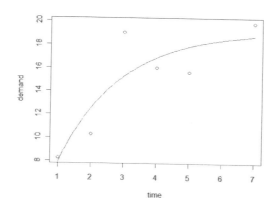

As noted in Eaves (1983), prior selection in nonlinear models is often carried out with an eye towards the resulting behavior of the posterior density. As with the Jeffreys prior, the selection of the prior is sensitive to the properties of the observed likelihood and the prior nature of the density selection is somewhat compromised. Typically in such models, prior selection is assumed to be independent for each parameter. Often non-informative improper priors are used, leaving the likelihood function itself to be normed and viewed as a posterior probability. But none of these may yield a stable posterior density. It depends on the degree of non-linearity in the model.

4.2.2   Example 5: Whale Population Dynamics Example

A likelihood model incorporating potentially chaotic behavior is examined in Givens and Poole (2002). A simple population dynamic model is developed for two key parameters in the estimation of the bowhead whale population. Data on abundance estimates for the years 1978–1988 and 1993 are generated using a simple nonlinear dynamic model for expected abundance, beginning in 1848. This provides the basic response data that is then used to support the observed likelihood. The structure of likelihoods in time series related data can be challenging as various parameters and the response may have autocorrelated structure.

Here the likelihood of the observed abundance N_t, and the structure of the underlying regression relationship are given by:

$$N_{1993} \sim Normal(\eta_{1993}, 626^2)$$

$$\log(N_{past}) \sim Normal_q(b\boldsymbol{\eta}, \widehat{\boldsymbol{\Sigma}})$$

$$\eta_{t+1} = \eta_t - C_t + 1.5(MSYR)\eta_t[1 - (\eta_t/K)^2]$$

where $MSYR$ is a productivity parameter, K is the equilibrium carrying capacity, C_t denotes the catch in year t, η_t is the expected total whale abundance in year t. The matrix $\widehat{\boldsymbol{\Sigma}}$ is the estimated correlation matrix for $\log(N_t)$ based on previously observed years and the scalar b is a fitting parameter given by 8293/7778. The unknown parameters in the model are $MSYR$ and K. The likelihood is set equal to zero if any non-positive simulated abundance arises.

4.3   Selecting Parametric Models and Likelihoods

The selection of models is often a challenging one. The goal is to integrate scientific understanding, mathematical and probability structures and the specific set of questions and formal hypotheses to be investigated. One important component in all models considered in this book is the choice of likelihood function. This is relevant to the second set of Questions in Chapter 1.

The likelihood function, in the context of modeling, is typically comprised of a mathematical relationship between outcome and explanatory variables nested within the probability structure. A useful practical distinction is categorical versus continuous scaling of the outcome variable.

Possible choices for statistical models typically fall in two overall classes: exponential and location-scale families of distributions. The normal distribution has a central place as an overlap of these wide ranging classes. Distributions from both classes have common application in many areas of research. While there are key differences in aspects of these classes from a frequentist perspective, for example in terms of conditional frequentist approaches to inference, from a Bayesian perspective these differences are not particularly useful.

The *location scale* form is generally given by:

$$\frac{1}{\sigma} f\left(\frac{y-\mu}{\sigma}\right)$$

where the mean is given by μ and variance by σ^2. Densities satisfying this form include the Normal, Student-t and modified Weibull. If we include a regression structure linking the independent response y_i for the ith subject with measured explanatory variables $X = [x_{1i}, x_{2i}, \ldots, x_{pi}]$ then we have as a location-scale model

$$\frac{1}{\sigma^n} \prod_{i=1}^{n} f\left(\frac{(y_i - \beta_0 - \beta_1 x_{1i} - \cdots - \beta_p x_{pi})}{\sigma}\right).$$

The *exponential family density* form with one-parameter, using a basic canonical parameterization, can be written:

$$f(y \mid \theta) = g(x)h(\theta) \cdot \exp\left[\phi(\theta)t(x)\right].$$

Densities with this form include the normal, exponential, binomial, poisson, Weibull, gamma and others. Much of standard statistics reflects this structure. With regression structure included, the resulting model is typically referred to as a *generalized linear model* and is often written:

$$\eta(y_i) = \beta_0 + \beta_1 x_{1i} + \cdots + \beta_p x_{pi} + \varepsilon_i$$
$$\varepsilon_i \sim f(\cdot)$$

Where the error components ε_i are assumed independent, $f(\cdot)$ is an exponential family density and $\eta(\cdot)$ (typically referred to as the *link function*) rescales the response y to allow for the explanatory variables on the right hand of the equation to be expressed on a linear scale.

The model can typically be fit to data using various possible combinations of link functions and parameterisations. The flexibility of these models and the likelihoods they generate are discussed in Lindsey (1997) and Dobson (1986). Some models are given in Tables 4.2 and 4.3.

Many other densities are available though these also typically fall within the exponential or location-scale families of distributions. Note that parameter meaning is important and will alter when transformations are applied to the response variable in these models. In real-world problems there may also be restrictions on parameter values (for example proportions adding up to one; $\sum_{i=1}^{n} \theta_i = 1$). Typically such restrictions are written directly into the model. These issues are illustrated in the examples considered in Part IV.

TABLE 4.2
Typical Continuous Data Models (Densities and Conjugate Priors)

	Density	Conjugate Prior
Normal known	$\left(\frac{1}{2\pi\sigma^2}\right)^{1/2}\exp\{-\frac{1}{2\sigma^2}(y-\mu)^2\}, y \in \blacksquare$	$p(\mu) = N(\mu_0, \sigma_0)$
Gamma	$\frac{\beta^\alpha}{\Gamma(\alpha)}\, y^{\alpha-1}\exp(-\beta y), y > 0$	$InverseGamma$
Exponential	$\theta e^{-\theta y}, \theta > 0, y > 0$	$Gamma(\alpha,\beta)$
Pareto	$\alpha\beta^\alpha y^{-(\alpha+1)}, y \geq \beta, \alpha > 0, \beta > 0$	$Pareto$
Multivariate Normal	$|\Sigma|^{-1/2}(2\pi)^{-k/2}\exp\{-(1/2)(y-\mu)'\Sigma^{-1}(y-\mu)\}$	$Normal(\mu,\Sigma)$

TABLE 4.3
Typical Count Data Models (densities and conjugate priors)

	Density	Conjugate Prior
Poisson	$\frac{\lambda^y e^{-\lambda}}{y!}, y = 0,1,2..$	$Gamma(\alpha,\beta)$
Binomial	$\binom{n}{y}p^y(1-p)^{n-y}, y = 0,1,2...n$	$Beta(\alpha,\beta)$
Negative-Binomial	$\binom{r+y-1}{r-1}\theta^r(1-\theta)^y, y = 0,1,2..$	$Beta(\alpha,\beta)$
Multinomial	$\frac{n!}{y_1!\cdots y_k!}p_1^{y_1}p_2^{y_2}\cdots p_k^{y_k},$ $\sum y_i = n, 0 \leq p_i \leq 1, \sum p_i = 1$	$Dirichlet(\theta_1,...,\theta_p)$
Geometric	$\theta(1-\theta)^y, y = 0,1,2..., 0 \leq \theta \leq 1$	$Beta(\alpha,\beta)$

4.4   Bibliography

[1] Bates D.M. and Watts D.G. (1988). *Nonlinear Regression Analysis and its Applications*. John Wiley and Sons, New York.

[2] Bernardo J.M. and Smith A.F.M. (1994). *Bayesian Theory*. John Wiley and Sons Inc. New York, NY.

[3] Boos D. D. and Monahan J. F. (1986), Bootstrap Methods Using Prior Information. *Biometrika* 73, No. 1, p. 77–83.

[4] Box G.E.P. and Tiao G.C. (1973). *Bayesian Inference in Statistical Analysis*. Addison-Wesley Publishing Company Inc. Reading, Massachusetts.

[5] Brimacombe M. (2014). Genomic Aggregation Effects and Simpson's Paradox. *Open Access Medical Statistics* 4, p. 1–6.

[6] Carlin B.P. and Louis T.A. (1996). *Bayes and Empirical Bayes Methods for Data Analysis*. Chapman & Hall, New York, NY.

[7] Carlin B. P. and Polson N. G. (1991). Inference for Nonconjugate Bayesian Models Using the Gibbs Sampler, *Canad. J. Statist.* 19, p. 399–405.

[8] Christensen R. (1987). *Plane Answers to Complex Questions*. Springer-Verlag New York Inc.

[9] Cohen J. (1988). *Statistical Power Analysis for the Behavioral Sciences* (2nd Ed) ISBN-13: 978-0805802832 ISBN-10: 0805802835. Lawrence Erlbaum Associates.

[10] de Finetti B. (1931). Funzione caratteristica di un fenomeno aleatorio. *Atti della R. Academia Nazionale dei Lincei, Serie 6. Memorie, Classe di Scienze Fisiche, Mathematice e Naturale*, 4:251–299.

[11] Dobson A.J. (1986). *An Introduction to Statistical Modeling*. Chapman & Hall Ltd., New York, NY.

[12] Draper N. R. and Smith H. (1981). Applied Regression Analysis, Second Edition, New York: John Wiley & Sons.

[13] Eaves D. M. (1983). On Bayesian Nonlinear Regression with an Enzyme Example, *Biometrika* 70, p. 373–9.

[14] Efron B. and Tibshirani R. (1994). *An Introduction to the Bootstrap*. Chapman & Hall/CRC. ISBN 978-0-412-04231-7.

[15] Fisher R.A. (1934). Probability, Likelihood and the Quantity of Information in the Logic of Uncertain Inference, *Proceedings of the Royal Society* A, 146, p. 1–8.

[16] Fisher R.A. (1955). Statistical Methods and Scientific Induction. *J. Roy. Statist. Soc. B* 17, p. 69–78.

[17] Garthwaite P.H., Kadane J.B. and O'Hagan A. (2005). Statistical Methods for Eliciting Probability Distributions. *J. Am. Statist. Assoc.* 100, p. 680–700. DOI 10.1198/016214505000000105.

[18] Gaudart J., Huiart L., Milligan P.J., Thiebaut R., Giorgi R. (2014). Reproducibility Issues in Science, is P value Really the Only Answer? *Proc Natl Acad Sci USA* 2014 May 13;111(19):E1934. doi: 10.1073/pnas.1323051111. Epub 2014 Apr 23.

[19] Gilks W.R., Richarson S. and Spiegelhalter D.J. (1996). *Markov Chain Monte Carlo in Practice*. Chapman & Hall, New York.

[20] Givens G. and Poole D. (2002). Problematic Likelihood Functions from Sensible Population Dynamics Models: A Case Study, *Ecological Modeling* 151, p. 109–124.

[21] Hicks C.R. (1982). *Fundamental Concepts in the Design of Experiments* *3rd ed.* Holt, Rinehart, Winston, New York.

[22] Jaynes E.T. (1957). Information Theory and Statistical Mechanics. *The Physical Review* 106, p. 620–630.

[23] Jeffreys H. (1934). Probability and Scientific Method, *Proceedings of the Royal Society* A 146, p. 9–16.

[24] Joliffe I.T. (1986). *Principal Component Analysis.* Springer-Verlag, New York.

[25] Julious S.A. and Mullee M.A. (1994). Confounding and Simpson's Paradox. *BMJ* 309, No. 6967, p. 1480–1481.

[26] Lindsey J.K. (1997). *Applying Generalized Linear Models.* Springer-Verlag, New York.

[27] Manly B.F.J. (1998). *Multivariate Statistical Methods, 2nd ed.* Chapman & Hall/CRC, New York, NY

[28] Miller Jr. R.G. (1997). *Beyond ANOVA: Basics of Applied Statistics.* Chapman & Hall/CRC, New York.

[29] Mosteller F. and Tukey J.W. (1977). *Data Analysis and Regression: A Second Course in Statistics.* Addison-Wesley, NY.

[30] Pearson K. (1901). On Lines and Planes of Closest Fit to Systems of Points is Space. *Philosophical Magazine* (6), 23, p. 559–572.

[31] Pham-Gia T. and Turkkan, N. (1992) Sample Size Determination in Bayesian Analysis. *Statistician* 41, p. 389–397.

[32] Sharpe D., (1997). Of Apples and Oranges, File Drawers and Garbage: Why Validity Issues in Meta-analysis Will Not Go Away. *Clin. Psych. Rev.* 17, December 1997, p. 881–901.

[33] Simpson E.H. (1951). The Interpretation of Interaction in Contingency Tables. *Journal of the Royal Statistical Society, Ser. B* 13: 238–241.

[34] Stone M. and Dawid A.P. (1972), Un-Bayesian Implications of Improper Bayes Inference in Routine Statistical Problems. *Biometrika* 59, p. 369–375.

[35] Tufte E.R. (1993). *The Visual Display of Quantitative Information.* Graphics Press, Cheshire CT.

[36] Wright S. (1921). Correlation and Causation. *J. Agricultural Research* 20:557–585.

4.5 Questions

1. The dataset in Example 2 can be examined using principal components analysis. A copy can be found in the data appendix to this

book. Obtain the first two principal components for the data. Plot them versus each other in a scatterplot and label them according to survival status. Argue that by using this rescaling of the data we obtain a picture of the physical scales related to natural selection in this bird population.

2. The Anscombe dataset is well known. A copy can be found in the data appendix to this book. Plot each dataset using a fitted simple regression line. Describe how the patterns in the data, all having similar regression lines, differ from each other. These examples show that assumed parametric models, even simple ones, require empirical and graphical investigation to be placed in their proper context and understood.

3. For one of the densities and corresponding likelihoods in Tables 4.2 and 4.3 respectively, derive the Jeffreys prior. Plot them as a function of the parameter(s) for a given generated dataset and discuss the weightings assigned to the various regions of possible parameter values.

4. Often a linear model is employed as a first step in developing a basic model for data. Develop a simple (x, y) set of data points that have a parabolic shape when scatterplotted. Is the correlation measure useful in the presence of non-linear patterns in the data? Why or why not? Does a log transformation linearize the pattern?

5. Using a standard package, for example R or Minitab, re-examine the fracking data in Example 1. A copy can be found in the data appendix to this book. Assume a Normal distribution. Find the m.l.e. for μ and σ. Plot the fitted probability model over the observed histogram for the data. Alter the bin size for the histogram. At what bin size does the unimodal Normal distribution no longer fit the data?

6. The literature of whale populations has been examined from both Bayesian and frequentist perspectives. See Givens and Poole (2002). Review relevant aspects of the literature and compare and contrast the two modeling approaches and their results with regard to predicting the whale population through time.

7. Review approaches to the development of prior densities in applications of multivariate linear regression for a fixed sample size n and a $Student - t$ error distribution. What is the Jeffreys prior in this context? Now assume we have a very large sample size. What is the asymptotic form of the Jeffreys prior? In general, as the amount of information underlying the likelihood increases, is the assumption of the non-informative Jeffreys prior more or less justified?

8. Examine the literature underlying the elucidation of prior expert belief. Discuss how such beliefs can be placed on a comparative

probability scale across experts. Develop a project where a group of experts are asked to formally assess their beliefs regarding a potential population characteristic (eg. proportion or rate of animals in a defined spatial area having a specific infection). Use a *Poisson* distribution for the likelihood with a generated sample of size $n = 12$ and a *Gamma* distribution as the prior density. This can be fit by eliciting mean and variance values. Plot the set of priors obtained and the respective posteriors. Discuss the variation observed in the prior densities and the stability of the posterior mean across the set of priors.

9. Review and discuss in detail the notion of hierarchy in the Bayesian setting. Since we can always technically write a joint prior density as the product of a suitable marginal and conditional density, we can always place Bayesian models in a hierarchical context, whether or not the likelihood itself reflects an underlying hierarchy or nested structure in relation to the sampling of observations. Select an example from the ecology or biology literature and discuss how the hierarchical sampling structure was represented in terms of an underlying mathematical and statistical model.

4.6 Suggested Readings

1. Box G.E.P. and Tiao G.C. (1973). *Bayesian Inference in Statistical Analysis*. Addison-Wesley Publishing Company Inc. Reading, Massachusetts.

2. Carlin B.P. and Louis T.A. (1996). *Bayes and Empirical Bayes Methods for Data Analysis*. Chapman & Hall, New York, NY.

3. Dobson A.J. (1986). *An Introduction to Statistical Modeling*. Chapman & Hall Ltd., New York, NY.

4. Edwards A.W.F. (1992). *Likelihood*. The Johns Hopkins University Press. Baltimore, Maryland.

5. Gelman A., Carlin J.B., Stern H.S., Rubin D.B. (1995). *Bayesian Data Analysis*. Chapman & Hall, New York, NY.

6. Gilks W.R., Richarson S. and Spiegelhalter D.J. (1996). *Markov Chain Monte Carlo in Practice*. Chapman & Hall, New York.

7. Hodges J.S. (1987). Assessing the Accuracy of Normal Approximations, *J. Amer. Statist. Assoc,* 82, p. 149–154.

8. Lange N., Ryan L., Billard L., Brillinger D., Conquest L., Greenhouse J. (1994). *Case Studies in Biometry*. John Wiley and Sons, New York.

9. Little R. J. (2006). Calibrated Bayes: A Bayes/Frequentist Roadmap. *J. Am. Statist. Assoc.* 60, p. 213–223.

10. O'Hagan A. (1994). *Kendall's Advanced Theory of Statistics, Volume 2B, Bayesian Inference.* John Wiley and Sons, New York.

11. Smith R.L. and Naylor J.C. (1987). A Comparison of Maximum Likelihood and Bayesian Estimators for the Three-Parameter Weibull Distribution. *Appl. Stat.* 36, p. 358–369.

Part III

Likelihood Based Statistical Theory and Methods: Frequentist and Bayesian

5

Introduction to Frequentist Likelihood Based Statistical Theory

5.1   Statistical Theory Related to Likelihoods

Statistical models involve the interpretation and application of probability on some scale. In both frequentist and Bayesian approaches, basic probability distributions are used to link observed data to the population characteristics of interest (parameters). Each applied discipline has a scientific literature supporting the relevance of specific probability models and providing guidance in their selection and application.

When parametric models are available and statistical inference is to be employed, this information is typically used to develop a likelihood function. In areas such as biology and ecology these may reflect very specific theoretical models. Once the likelihood function is developed, the frequentist and Bayesian approaches differ in how they extract relevant information from the likelihood function and how this information is employed for statistical inference. In some complex settings, often with measurements through time, the likelihood function may be intractable. Here we assume the likelihood function can be determined.

The overlap and agreement between frequentist likelihood based methods and the Bayesian approach is often substantial, as they share the foundation of the model-data combination expressed through the likelihood function, though examples exist where they differ even when assuming very little (David et al.,1973). In general, Bayesian methods often allow greater flexibility in estimation and modeling procedures, subject to the assumption of additional prior assumptions.

Note that technically the local shape of the log-likelihood function underlies much statistical theory, frequentist and Bayesian. This dependence is reflected in recent advances such as likelihood-based frequentist saddlepoint approximations (Fraser et al., 1999) and (Bayesian) Laplace approximations (Tierney and Kadane, 1986). Further reference to the likelihood function and related application of Bayesian methods can be found for example in Berger and Wolpert (1988) where the likelihood principle, the idea that only aspects of the model-data setting of direct relevance to the likelihood need be considered for inference, is examined in detail.

While frequentist and Bayesian methods share the foundation of likelihood, the subjective nature of Bayesian probability represents a distinct utilitarian based perspective on the use of probability. Technically related to frequentist-likelihood approaches, the interpretation of Bayesian probability is slightly different, representing a learning oriented approach to the assessment of model and data. From a baseline of prior belief regarding θ, the model-data based likelihood function is employed, along with Bayes theorem to obtain a set of updated beliefs regarding θ in the form of the posterior density.

5.1.1   Example: Normal Distribution

As both Bayesian and frequentist approaches can be sensitive to large sample arguments, it is worthwhile to review briefly the likelihood function arising in the case of the normal distribution. Consider a sample of n individual measurements y_i, collected independently from a common population. If we assume that the probability distribution describing the values y_i follows a normal distribution with mean μ and known variance σ_0^2, which we write as $y_i \sim N(\mu, \sigma_0^2)$, then the likelihood function for μ resulting from this model is given by

$$L(\mu \mid y_1, \ldots, y_n) = c \cdot \exp\left[-\frac{1}{2}\frac{(\mu - \overline{y})^2}{\sigma_0^2/n}\right]$$

which, as a function of μ, actually follows the general shape of a normal probability distribution centered at $\overline{y}$.

In most larger samples, the likelihood function, on some scale, will tend towards a normal shape (the parameters may have to be rescaled). Exact and accurate approximate methods now exist to support frequentist and Bayesian calculations for many different likelihood functions and corresponding priors, so there is no immediate need for assumptions of normality in small samples. But in larger samples, where the central limit theorem affects the statistics underlying the likelihood function, the likelihood will tend towards the normal shape and induce a growing similarity between frequentist and Bayesian results. See Sprott (1980). This is especially true for point estimation. The approximate Bayesian calculation (ABC) method reflects this very directly. See for example Aeschbacher et al. (2012).

5.2   Basic Statistical Models

We now review some of the statistical models and methods that form the basis of much standard statistical analysis. All students of statistical methods should know them and view them as tools for understanding patterns in data, in the presence of random error, whether the final context is Bayesian or frequentist. A review of some of the history of these methods can be found in Stigler (1986) and Box (1978). All these models also generate or reflect underlying likelihood functions and therefore can also be placed in a Bayesian context when desired. Bayesian extensions and applications of several of these models are presented and applied in Part IV.

5.2.1   T-Test

It is useful to note that test statistics typically correspond to a specific under-lying probability model. The (Student's) t-test, first suggested by Gossett, a brewer who wished to better assess aspects of the brewing process, and later placed in a formal mathematical context by R.A. Fisher, typically allows for (i) comparison of a single mean $\bar{y}$ to a specific hypothesized mean value μ and (ii) comparison of two means, for example the mean difference between a treatment and control group.

One-sample

In one dimension the basic statistic, generally referred to as a pivotal quantity in the sense that its standardized sampling distribution will be the basis for statistical inference, is given by:

$$\frac{(\bar{y} - \mu)}{s/\sqrt{n}}.$$

This can be seen as corresponding to the following model: consider a sample of n individual measurements y_i, collected independently from a common population with probability distribution following a normal distribution with mean μ and unknown variance σ^2. We write this as $y_i \sim N(\mu, \sigma^2)$ and look to test the hypothesis $H_0 : \mu = \mu_0$. To do this, we assume the true value of the mean parameter is indeed μ_0 and look for evidence that this hypothesis is incorrect.

We use the sampling theory result:

$$T_1 = \frac{(\bar{y} - \mu_0)}{s/\sqrt{n}} \sim t_{n-1}$$

where t_{n-1} denotes a Student-t distribution with $(n-1)$ degrees of freedom. Note that in larger samples $(n > 30)$, the t_{n-1} distribution will approach the $N(0,1)$ distribution.

We carry out this testing by looking at the pivotal quantity carefully. In a direct sense it measures the number of standard errors $(s/\sqrt{n})$ that lie between the sample mean and hypothesized population mean value $(\bar{y} - \mu_0)$. If this difference is large, typically 3 or higher, we reject the null hypothesis; ie. it is not supported by the data observed in the context of the assumed model and data generating process. To assess this difference using a probabilistic scale we can look at the p-value or tail area:

$$p - value = P(|T_1| > T_{obs})$$

and reject the null hypothesis if the p-value is small, as this will correspond to a large difference in standard errors. Note that here we are assuming an alternative of not equal, thus the use of absolute value.

Defining what is meant by a "small p-value" can be a challenge. Fisher suggested using a value of one in twenty (0.05) as a useful cutoff point, but was flexible in this regard. Note if the null hypothesis is actually correct and we simply observe a rare sample, then the p-value can be interpreted as an error rate. This typically justifies a 0.05 or "one in twenty" cut-off.

This general approach to the testing of hypotheses is in accord with the way in which scientific inference has been taught for a very long time and corresponds to the philosophy underlying frequentist hypothesis testing. We prove nothing directly. Rather we assume that what we are testing is true and look to refute the assumption. David Hume, Karl Popper and other philosophers of science have pointed out the need to prove results via contradiction, as future samples are not yet known. It is the basis of a long debate both in the philosophy of science and within statistics itself (Stigler, 1986).

Note that Bayesian approaches to hypothesis testing do not assume the null hypothesis is true. Rather we compare the relative weight of competing values for the parameter of interest using the posterior density. This is further discussed below.

Two-sample

In two dimensions, the t-test pivotal statistic is given by

$$T_2 = \frac{(\overline{y_1} - \overline{y_2}) - (\mu_1 - \mu_2)}{s_{pooled}}$$

$$s_{pooled}^2 = \frac{(n_1 - 1)s_1^2 + (n_2 - 1)s_2^2}{n_1 + n_2 - 2}$$

where s_1^2 and s_2^2 are the sample variances from each sample respectively.

This corresponds to a basic model; consider two samples of n_1 and n_2 individual measurements respectively and denote the respective observations y_{1i} and y_{2i}. Assume these are collected independently, each respectively from a population with probability distribution $y_{ij} \sim N(\mu_i, \sigma^2)$, $i = 1, 2$ and $j = 1, ..., n_i$. We want to test the hypothesis of equality, $H_0 : \mu_1 - \mu_2 = 0$.

In this case, assuming the null hypothesis is true, the relevant sampling theory result is:

$$T_2 = \frac{(\overline{y_1} - \overline{y_2})}{s_{pooled}} \sim t_{n_1+n_2-2}$$

and the p-value $= P(|T_2| > T_{obs})$ the area at or beyond the observed value of T where an alternative of not equal is assumed. If the p-value < 0.05, we reject the null hypothesis; the data does not provide sufficient evidence to support the null hypothesis of equivalent mean values for the two populations being sampled.

Example: Bird Survival Data

Looking at each measurement variable and comparing the values for survival versus non-survival groups using two sample t-tests gives the following:

TABLE 5.1
Bird Survival Data T-Tests

Variable	Total Length	Alar	Beak and Head	Humerus	Keel of Sternum
p-value	0.315	0.688	0.843	0.731	0.914

While none of the individual variables shows a significant difference between the groups, correlation based patterns are more insightful. Note that since we are conducting a set of five hypothesis tests on the same core dataset it is typically viewed as necessary to adjust the Type I error cutoff here from 0.05 to $.05/5 = 0.01$. The so-called Bonferroni adjustment to maintain an overall family Type I error of 0.05.

5.2.2   ANOVA

In the broader setting where there are more than two groups to be compared, R.A. Fisher developed a generalization of the t-test called the Analysis of Variance (ANOVA). In the case of 1-way ANOVA, we have k groups and are testing the null hypothesis $H_0 : \mu_1 = \mu_2 = \cdots = \mu_k$. This is called one-way ANOVA as the k groups to be compared typically correspond to k levels of a single factor, for example exposure levels of a given chemical or different levels of a temperature sensitive variable.

To test the null hypothesis the following model is assumed:

$$y_{ij} = \alpha + \mu_j + \varepsilon_{ij} \tag{5.1}$$

$$\varepsilon_{ij} \sim N(0, \sigma^2) \tag{5.2}$$

where y_{ij} is the response of the ith individual in the jth group, α is a baseline level, μ_j the average level of the jth treatment and the ε_{ij} are errors, assumed independent and uncorrelated with the levels of the factor in question. The variation in the problem is assumed to be similar or homogeneous in all k levels and modeled by a single parameter σ. This may require initial rescaling of the outcome variable, for example taking logarithms.

The pivotal quantity used here is typically the F-statistic, named for R.A. Fisher and given by:

$$F = \frac{\sum(\bar{y}_j - \bar{y})^2/(k-1)}{\sum(y_{ij} - \bar{y}_j)^2/(n-k)} \sim F_{k-1, n-k} \tag{5.3}$$

where $\bar{y}$ is the overall average and $\bar{y}_j$ is the average in the jth group and $n = \sum n_j$. This statistic has a $F_{k-1,n-k}$ distribution when the null hypothesis is true. The p-value is then given by $P(F_{k-1,n-k} > F_{obs})$. A small p-value implies rejection of the null hypothesis.

It is worth noting that Fisher, when applying the F-statistic typically used $(1/2)\ln F$ as the test statistic rather than F itself. This simple $1 - 1$ transformation has a more symmetric distribution allowing for a more direct interpretation of the observed value of the pivotal quantity in terms of standard error. Note that for both the t-test and F-test, the assumption of normality of the errors is necessary or a larger sample supporting application of central limit theorems.

The corresponding standard ANOVA table, which decomposes overall variation $SST = \sum(y_{ij} - \bar{y})^2$ into the variation of individual data points about the group mean, $SSE = \sum(y_{ij} - \bar{y}_j)^2$, plus the variation of the group means themselves about the overall mean $SSR = \sum(\bar{y}_j - \bar{y})^2$, reflects an elegant underlying geometry of orthogonal projections (these sum of squares are the squared lengths of such projections and have independent chi-square distributions when the original data is normally distributed).

ANOVA methods can be extended in many directions including cases of many factors, each with several levels and potential interactions between the levels. This tends to go under the name Design of Experiments. See for example Hicks (1982) for an overview. It is sometimes the case that this type of detailed model information and decomposition of sources of variation can be lost when applying Bayesian methods where averaging over the parameter space directly may mask such detailed model breakdowns. This type of consideration supports a more integrated and broad based approach to statistical analysis.

Example Consider data collected for a response y_{ij} across three treatment groups, one of which is a control group.

A scatterplot for the data and the ANOVA table accompanying this graphic is given below. There is an overall difference among group means as shown by the F-test. The adjusted pairwise t-tests at the 5% level give significant differences between Controls and Group1, Controls and Group2, and Group1 and Group2.

TABLE 5.2
Data for 1-Way ANOVA

Ctrls	10	11	12	9	8	10	12	13	12	13	11	14	13	12	11	11	10
Gp1	20	21	22	23	22	21	19	18	23	22	20	17	18	19	20	24	23
Gp2	40	38	42	41	44	43	42	41	40	41	42	37	44	43	38	39	40

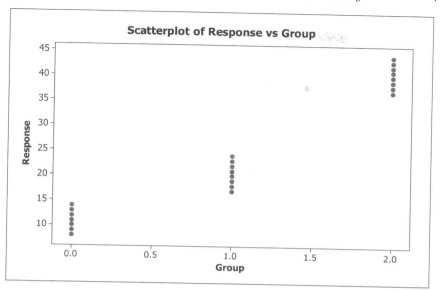

FIGURE 5.1
Data Scatterplot

TABLE 5.3
1-Way ANOVA

Source	df	SS	MS	F	p-value
Factor	2	7769.76	3884.88	1054.58	0.0001
Error	48	176.82	3.68		
Total	50	7946.59			

5.2.3 More on Linear Models

Linear models are often employed in settings where underlying relationships among variables are to be investigated. While they typically serve as an automatic first choice for researchers, the determination of scales upon which linear relationships exist is an important first step in the analysis of biological and ecological data. This is often carried out graphically as discussed in Chapter 2. Linear models are "linear" only in the degree of the parameters themselves, so they can actually be fairly flexible, for example employing $x_1 = \sin(x)$ and $x_2 = \cos(x)$ waves as explanatory variables, and still meet the criteria to technically be linear models.

Linear models are typically written in vector notation as $y = X\beta + \varepsilon$ which can be re-expressed as:

$$y_i = \beta_0 + \sum_{j=1}^{p} \beta_j x_{ij} + \varepsilon_i \qquad (5.4)$$

where y_i is the response of interest collected from the ith individual, ε_i is the associated random error, typically independent and normally distributed (not required in larger samples) with mean 0 and common variance σ^2, $X = [x_1, x_2, ..., x_p]$ the design matrix with x_j column vectors each representing $i = 1..., n$ measurements on the jth explanatory variable.

Frequentist inference for specific β_j parameters can be carried out using least squares or maximum likelihood estimation and an ANOVA format to develop the relevant test statistics. Computer based resampling via application of the bootstrap procedure can also be used to determine significance. These tend to be based on variations of the F-statistic; the general linear F-test or partial F-statistics, measuring the extent to which overall relative variation is better explained by adding the jth explanatory variable (or a combination thereof) in the regression equation. See for example Draper and Smith (1998). This is discussed in more detail in Chapter 4.

Linear models also have a goodness of fit measure available; the R^2 value which can be interpreted as comparing the fit of the data under two competing linear models **A** and **B**:

$$\textbf{A. } y_i = \beta_0 + \varepsilon_i$$
$$\textbf{B. } y_i = \beta_0 + x_{1i}\beta_1 + x_{2i}\beta_2 + \cdots + x_{pi}\beta_p + \varepsilon_i$$

In each model, the error components ε_i are assumed independently distributed (again, typically normal, but not required in larger samples) with mean 0 and a common variance σ^2.

In Model **A** the least squares estimate of β_0 is given by the average of the response values, $\bar{y}$. The resulting residual sum of squared errors,

$$\textbf{SST} = \sum (y_i - \bar{y})^2$$

can also be seen as an estimate of total variation in the response without reference to the x_{ij} explanatory variables; a measure of underlying background noise. Inclusion of explanatory variables should result in a smaller residual error if the explanatory variables selected by the researcher are indeed related to the response.

In Model **B** the least squares estimates $\widehat{\beta}_0, \widehat{\beta}_1, \ldots, \widehat{\beta}_p$ for $\beta_0, \beta_1, \ldots, \beta_p$ are obtained by applying least squares. The residual sum of squared errors is then defined by:

$$\textbf{SSE} = \sum_{i=1}^{n} (y_i - \widehat{\beta}_0 - x_{1i}\widehat{\beta}_1 - x_{2i}\widehat{\beta}_2 - \cdots - x_{pi}\widehat{\beta}_p)^2$$

and the R^2 value measures the relative decrease in unexplained variation or percent improvement in fit as the researcher moves from Model **A** to Model **B**:

$$R^2 = (\mathbf{SST} - \mathbf{SSE})/\mathbf{SST}$$

The R^2 value lies between 0 and 1. It is expected that if the researcher has chosen wisely, the **SSE** value (variation about the linear model) will be substantially less than the **SST** value (baseline variation in the response without the linear model). The closer the R^2 value lies to 1, the better the fit of the model. Note that due to the geometry of the linear model, if the constant β_0 is left out of the model the R^2 value may be negative.

Example Consider the following dataset comprising responses y and x.

TABLE 5.4

Data for Linear Regression

y	3	5	7	9	12	13	11	15	17	18	21	22	23	22	21
x	1	2	2	3	4	5	4	6	7	5	6	8	9	7	8

The data and fitted regression line are shown. The regression line is given by $\widehat{y} = 1.309 + 2.589x$ and the goodness of fit is high ($R^2 = 91\%$). The fitted slope 2.589 implies that the response y will increase on average by 2.589 units for a 1 unit increase in x.

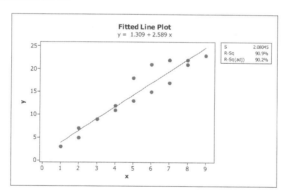

FIGURE 5.2

Fitted Regression Line

5.2.4 Centering and Interaction Effects in Linear Models

The basic structure of a linear model can be extended and modified by using transformations of scale such as the log transformation, the centering of variables (subtracting off the respective mean from each variable in the model) and the inclusion of interaction terms $(x_{1i} \cdot x_{2i})$. These terms are added into the model by the researcher as appropriate, giving the linear model surprising flexibility. However in such settings the model must be interpreted carefully. A fitted linear model with many highly correlated explanatory variables implies that relationships between the response y and each individual explanatory variable x_j may not represent the entire story regarding the relationship of y to the set of explanatory variables $x_1, ..., x_p$ which are sometimes best viewed as a single group of variables.

It is interesting to investigate and clarify centering effects in this setting, especially in regard to models with interaction terms. To begin, the standard linear model can be expressed in the form:

$$y_i = \beta_0 + \beta_1 x_{1i} + ... + \beta_p x_{pi} + \varepsilon_i.$$

Assume the ε_i are *i.i.d* $N(0, \sigma^2)$ random errors, responses of interest y_i, linear parameters β_j, and explanatory (fixed) variables x_{ij}. Using vector and matrix notation this can be re-expressed as $y = X\beta + \varepsilon$. See for example Draper and Smith (1981).

The use of centering in linear regression is often justified as a way to lower the observed correlation among the explanatory variables. For example, if x_{ij}^2 is placed in a model containing x_{ij} then centering will often reduce the correlation between them. This improves the stability of the fitted model. See for example Echambadi and Hess (2007).

The vector of values $x_{ij} \cdot x_{ik}$ in a linear model is usually taken to represent interaction effects as the partial derivative of the response y with regard to x_j when including this term in the model will have the general form:

$$\frac{\partial y}{\partial x_j} = \beta_i + \beta_j x_k$$

which implies the main effect of x_j on the response y is dependent on the level of x_k. Centering is often thought to be useful when interaction terms are included in a linear model, again to improve stability and interpretability in least squares based estimation.

Without centering it is sometimes argued that cross-product terms used in regression models to model interaction effects may be highly correlated with main effects, making it difficult to identify main and interaction effects respectively. However in such models centering does not alter the sampling accuracy of the main effect parameter estimates, the values of the interaction effect estimates, nor alter the overall R^2 goodness of fit. Note that the test statistics for main effects may require adjustment as the respective parameters in the model may slightly alter meaning after centering the variables. All that centering achieves is the possible lowering of the correlations among the interactions terms themselves, which allows for clearer interpretation of these effects individually.

To see this, first consider the simple regression model:

$$y_i = \beta_0 + \beta_1 x_i + \varepsilon_i$$

and assume this is used to model the linear pattern in an (x, y) data scatterplot. Centering by definition will not affect the shape of the initial (x, y) data cloud, it simply re-centers it to $(0,0)$. The best fitting line through the data cloud will therefore not alter in its slope nor in its residuals. As the sums of squares of the error SSE is the squared length of the residual vector, the average squared length (MSE) and goodness of fit measure $R^2 = 1 - SSE/SST$, where $SST = \sum(y_i - \bar{y})^2$, also do not alter with centering.

The least squares estimate for the slope, $\widehat{\beta}_1$ is based on sums of differences from the x and y means and is therefore invariant to centering, as is the correlation between x and y. Note that the chosen error distribution does not affect these results which reflect the geometric properties of the least squares estimators. The estimate for the intercept β_0 will alter after centering the data.

For the multivariate linear model:

$$y_i = \beta_0 + \beta_1 x_{1i} + ... + \beta_p x_{pi} + \varepsilon_i$$

the same basic geometric argument holds in regard to residuals and gives similar results in relation to centering the data. The centering of all variables in this model has no effect on measures of association between the x_j and y variable pairs, including least squares estimators $\widehat{\beta}_j$, $j = 1, ..., p$. Note that if second-order terms $\beta_j x_j^2$ are added to the model, centering may lower the correlation between x_j and x_j^2 terms.

The addition of interaction terms $x_j \cdot x_k$ to this model can generally be used to examine whether the relationship between y and x_j is sensitive to the levels of another variable x_k. It is generally accepted that if the coefficient for the respective interaction term is found to be significant, the main effect relating y and x_j cannot be directly interpreted and stratification of the model and data may be necessary to interpret observed results.

As noted above centering will not alter most overall measures of fit for the linear model, even in linear models where some of the variables are interaction terms. In particular, if we have as our model:

$$y_{ij} = \beta_0 + \beta_1 x_{1i} + \beta_2 x_{2i} + \beta x_{1i} x_{2i} + \varepsilon_{ij}$$

the least squares estimator for the interaction term $x_1 \cdot x_2$ will not alter if both the x_1 and x_2 variables are centered, neither will the R^2 value for the model. The significance of the main effects here will appear to alter, but only due to the parameters having a different meaning in the centered model and the related t-tests testing slightly different hypotheses. All that really occurs is that the interactions terms themselves become less correlated. See Brimacombe (2016) and the references therein.

5.2.4.1   Example: Penrose Bodyfat

The Penrose bodyfat (Penrose et al., 1985) dataset of physiologic measurements contains several variables that are highly correlated. Let bodyfat density (y_i) be the response of interest in relation to several body measurements (x_{ij}); Abdomen, Ankle, Biceps, Chest, Forearm, Knee, Hip, Thigh, Weight, Wrist. Linear regression analysis gives three significant variables (Abdomen, Weight, Wrist) accounting for an R^2 value of 73.1%. These variables themselves are highly correlated (.88, .73, .62) and these values do not alter if the measurements are centered.

If interactions are then included in the model, dropping the Abdomen-Weight interaction due to extreme collinearity, the correlations among the interaction terms pre-centering (.95, .96, .94) and post-centering (.38, .90, .30) show effects of centering. The overall F-test value of 133.95 (significant at 0.0001) and $SSE = 0.02$ are shown to be invariant to centering as is the R^2 value. The least squares estimates for the interactions terms and their standard errors do not alter. These results are given in Table 5.1.

Nonlinear Models

It is interesting to note that the effects of centering are somewhat dependent on the linearity of the model. If the model is nonlinear and second order approximations are used in tandem with local linear approximations then centering must be interpreted carefully. See Brimacombe (2016) for detailed discussion.

5.2.5   High-Dimensional Linear Models

A standard tool for understanding data and linking a response variable to a set of explanatory variables is the linear model:

$$y = X\beta + \varepsilon$$

where y is a $(n \text{ x } 1)$ vector of responses, $X = [x_1, ..., x_n]$ a $(n \text{ x } p)$ matrix of p measured variables, and ε a $(n \text{ x } 1)$ vector of error components. If all variables x_i are thought to be relevant, the fitted least squares model is given by $\hat{y} = X\hat{\beta}$ where $\hat{\beta} = (X'X)^{-1}X'y$.

When many variables x_i have been collected and only a few are believed to be relevant to predicting y, placing a restriction on the model, for example allowing only a few of the β_i values to differ significantly from zero, can help find the "best" underlying linear model, most typically using stepwise methods. A review can be found in Brimacombe (2017).

As noted earlier, the usual sparseness restriction in these settings can be expressed:

$$\sum_{i=1}^{k} |\beta_i|^m < t$$

TABLE 5.5
Centering Effects in Linear Models with Interaction

Variables	Weight	Abdomen	Wrist	Abdomen*Weight	Abdomen*Wrist
Original Data $(s = 0.0099; R^2 = 73.1)$					
Coeff(Std Error)	-.0005(.00041)	-.0026(.00041)	-.0017(.0056)	.000002(.000002)	.00003(.00003)
p-value	0.18	0.001	0.77	0.29	0.35
Centered Data $(s = 0.0099; R^2 = 73.1)$					
Coeff(Std Error)	.0002(.00005)	-.0022(.00013)	.0037(.001)	.000002(.000002)	.00003(.00003)
p-value	0.001	0.0001	0.001	0.29	0.35

for small values of t and k. Sparseness restrictions usually assume $m = 1$ or 2 and allow for identifiable parameter estimation when the linear model suffers from less than full rank in the design matrix X and the least squares estimator $\widehat{\beta} = (X'X)^{-1}X'y$ does not exist.

Note further that in regard to the parameter space, such restrictions limit the set of possible β_i combinations that may be examined. This implicitly affects any assumed prior densities from the Bayesian perspective and the sparseness restriction is sometimes incorporated into the definition of the prior density. The shape of this restricted parameter space depends on m. From the perspective of the sample space sparseness limits potential values for the estimators $\widehat{\beta}_i$ allowing for the application of the usual least squares approach and identifiable parameter estimates. See Brimacombe (2017).

5.2.5.1   Ridge Regression

When there are highly correlated variables in the X design matrix these affect the stability of $(X'X)^{-1}$ and the ridge regression approach can be applied. This uses a sparsity restriction with $m = 2$. It is easy to examine the ridge regression based estimator when $n > p$. Assuming centered data, we can write this as:

$$\widehat{\beta}^R = (X'X + \lambda I)^{-1}X'y$$

for some scalar λ. Even with high correlation affecting the X design matrix this estimator will exist. The value for λ can be chosen graphically or using an assumed prior density for λ and Bayesian posterior calculation.

The Singular Value Decomposition (SVD) method can help to better understand this estimator. To apply SVD to the X matrix we write:

$$X = UDV'$$

where $U = (u_1, ..., u_p)$ is a n by p orthogonal matrix and the u_j form an orthonormal basis for the column space of X, V is similarly constructed orthogonal matrix for the row space of X and D is a diagonal matrix $(d_1, ..., d_p)$.

A geometric perspective can be applied here to better understand the approach. Here ridge regression is attained by projecting y onto the normalized principal components of X (ie. the u_j vectors) where the jth principal component of X is given by $d_j u_j$. So the ridge regression estimator can be re-expressed as:

$$\widehat{\beta}_j^R = \frac{d_j^2}{d_j^2 + \lambda} u_j' y$$

which is a weighted projection of y onto the principal component u_j using the relative weights of d_j and λ. Eigenvalues and related principal components play an important role in the theory and interpretation of linear models and data matrices. For more discussion of related approaches in high dimensional data settings, see Brimacombe (2014).

5.2.6 Generalized Linear Models

In cases where the error distribution does not follow the Normal distribution, but follows a distribution in the exponential family of distributions, the linear model format can be extended to include for example Poisson regression (count data within a specific time period, disease rates), logistic regression (binomial outcome), Weibull regression (survival and reliability) and others. These are typically referred to as examples of generalized linear models. References include Dobson (1986), McCullagh and Nelder (1989). These were reviewed in Chapter 2.

Linear Model with Normal Errors

For the Normal case the relevant model for an *i.i.d.* sample of size n is given by:

$$y_i = \beta_0 + x_{1i}\beta_1 + x_{2i}\beta_2 + \cdots + x_{pi}\beta_p + \varepsilon_i = X\beta + \varepsilon$$
$$\varepsilon_i \sim N(0, \sigma^2)$$

where $i = 1, ..., n$, $X = (x_1, x_2, ..., x_p)$, an array of column vectors, one for each explanatory variable, $\beta' = (\beta_0, \beta_1, \beta_2, ..., \beta_p)$ and $\varepsilon' = (\varepsilon_1, ..., \varepsilon_n)$. Assume the errors are independent

The corresponding likelihood can be written in matrix notation:

$$L(\beta, \sigma^2 | data) = c \cdot \frac{1}{\sigma^n} \exp\left[\frac{-1}{2\sigma^2}(y - X\beta)'(y - X\beta)\right].$$

This is the basis of much frequentist maximum likelihood based inference and Bayesian methods in linear models.

Logistic Regression Model

The logistic regression model is a commonly applied regression model. The distribution of the response variable is binomial (a binary outcome is the response, for example success/failure or above/below a given threshold) and the probability of success can be directly modeled. The likelihood function is initially written:

$$L(\boldsymbol{\pi} \mid \mathbf{y}) = c \cdot \prod_{i=1}^{n} \pi^{y_i} \cdot (1 - \pi)^{1-y_i} \tag{5.5}$$

where y_i is a $0-1$ binary response variable denoting success at the x_i setting, π is the probability of a success for each individual and the n individual binary responses are assumed to be independent Bernoulli random variables. The probability of a success at the x_i setting is then modeled and assumed to follow a logistic function form:

$$\pi = \pi(x_i) = \frac{\exp(\sum_{j=0}^{k} \theta_j x_{ij})}{1 + \exp(\sum_{j=0}^{k} \theta_j x_{ij})}$$

where $x_i = (x_{i0}, ..., x_{ik})'$ denotes the ith setting of the explanatory variables,

θ_0 is the constant in the model and $x_{i0} = 1$. This is then combined with (5.5) to yield a likelihood function for $\boldsymbol{\theta} = (\theta_0, \theta_1 \ldots, \theta_k)$.

$$L(\boldsymbol{\theta} \mid \mathbf{y}) = c \cdot \prod_{i=1}^{n} \left[\frac{\exp(\sum_{j=0}^{k} \theta_j x_{ij})}{1 + \exp(\sum_{j=0}^{k} \theta_j x_{ij})} \right]^{y_i} \left[1 - \frac{\exp(\sum_{j=0}^{k} \theta_j x_{ij})}{1 + \exp(\sum_{j=0}^{k} \theta_j x_{ij})} \right]^{1-y_i}$$

While this is often a complicated nonlinear function, iteratively reweighted least squares, usually in the form of Fisher's scoring method, can be applied to obtain maximum likelihood estimates with associated random errors for $\boldsymbol{\theta}$. These calculations are standard in most statistical packages. Convergence is not however guaranteed due to the nonlinearity involved in the likelihood function and inference may be unstable if there are small cell counts in some sub-groups of the variables. See Hauck and Donner (1977). McCullagh and Nelder (1989) provide a detailed discussion of the generalized linear model.

5.2.7   Random Effects

Random effects, has a long history and is typically analyzed using the variance components approach of ANOVA. It can be viewed as a partial Bayes approach imposed by the design of the study in question. For example, if we wish to examine the relationship between the average size of a species and the abundance of this species, but only have the resources to sample at a small number of sites chosen at random from a longer list of possible sites, we might allow the average size value at a given site, μ_k, to have its own random distribution, typically the normal distribution. A similar argument holds where we study a rare human disease using several medical centers to collect data.

This is often carried out in the context of frequentist inference using linear models leading to the title "mixed models" and generalizes the repeated measures setting. The linear random effects model can be written:

$$y_{ij} = X_{ij}'\beta + Z_{ij}'\gamma_i + \varepsilon_i$$

where X_{ij} is a vector of explanatory variable values for the ith subject, Z_{ij} is a similar design vector, $\varepsilon_i \sim i.i.d. N(0, V_i)$, and $\gamma_i \sim N(0, G)$ with γ_i and ε_i also assumed independent. See for example Laird and Ware (1982). These methods can also be written as extensions of generalized linear models, especially when reflecting time dependent data. In the fully Bayesian context random effects is best approached as an example of the hierarchical modeling approach and is discussed below.

Random effects also has natural application in longitudinal or repeated measures settings, where the set of randomly selected individuals being subject to follow up are a sample of a much larger population. Baseline measurements on each subject may vary and require a distributional assumption. Genetic studies are also an area of application as the set of individuals examined are only a small component of the overall population being studied and may cluster in terms of their genes due to family or other underlying structures, both sampling related and genetic in nature.

5.2.8 Nonlinear Models

Many biological and ecological phenomena and relationships are nonlinear in nature. While transformations can be used to make some scales more linear, researchers often prefer to use more realistic nonlinear models. Processes that involve growth over time, dose-response assessment, and ecological monitoring of weather related variables are examples. To model some of these we can employ nonlinear regression models.

The standard nonlinear regression model and associated likelihood function can be written:

$$y_i = \eta(x_i, \boldsymbol{\theta}) + \varepsilon_i$$

$$L(\theta, \sigma^2 | y_i, x_i) = c \cdot \frac{1}{\sigma^n} \exp[\frac{-1}{2\sigma^2}(y_i - \eta(x_i, \boldsymbol{\theta}))'(y_i - \eta(x_i, \boldsymbol{\theta}))]$$

$i = 1, \ldots, n$ where x_i are fixed values of the explanatory variable x, the model function $\eta(\cdot)$ is of known form and depends on the parameter vector $\boldsymbol{\theta} \in \mathbf{R}^p$ and the x_i. The ε_i are independent error terms, each normally distributed $N(0, \sigma^2)$. This can also be extended to generalized linear models where the error distribution is non-normal. The commonly used logistic regression is technically a type of nonlinear regression as are most generalized linear models.

The properties of the likelihood function, or likelihood surface (more than one β parameter) in nonlinear settings affect the stability of the usual likelihood based test statistics. In fact some likelihood related test statistics, such as the Wald statistic, become potentially unstable in a nonlinear model setting, while the likelihood ratio test statistic can yield inference regions (typically elliptically shaped in linear models) which are difficult to interpret. This often leads to the need for re-scaling of parameters, for example writing the model in terms of $\phi = \ln(\theta)$ rather than the original θ parameterization. See for example Seber and Wild (1989) and Pawitan (2000). The shape of the likelihood about its mode is a key component of both frequentist and Bayesian inference. A plot of the log-likelihood function for the BOD data and associated nonlinear regression model given in Chapter 1 is shown.

Note that from the Bayesian perspective use of nonlinear models results in both more complex posterior densities and issues regarding the selection of priors, arising often from the need to reparameterize the model. In general, if $p(\theta)$ is a uniform prior for θ and we change variables $\phi = g(\theta)$, the resulting prior for ϕ is proportional to

$$p(\phi) = | \partial g^{-1}(\phi)/\partial \phi | \tag{5.6}$$

Log-Likelihood Plot for BOD Model and Data

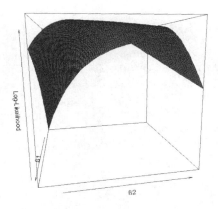

FIGURE 5.3
Log-Likelihood Plot for BOD Model and Data

and this will typically not be uniform in the original ϕ parameter. R.A. Fisher viewed this as a serious difficulty, nonlinearities being common in biological and genetic models. Simply changing variables will be informative as rescaling the parameters will alter the shape of the prior due to the effect of changing variables to ensure the area under the prior remains equal to one. In this way a "flat" or non-informative prior may become informative.

To deal with this difficulty, thoughtful application of Bayesian methods in non-linear models will initially reparameterize the likelihood such that the new parameter ϕ is the parameter of interest and the resulting likelihood is fairly stable. A non-informative prior is then selected for ϕ. The $\phi = g(\theta)$ transformation however needs to be interpretable within the context of model and science. Eaves (1983) suggests use of

$$\frac{|F'(\theta)F(\theta)|}{\sigma}$$

where $F(\theta)$ is the tangent model at θ to the original non-linear regression surface or solution locus. This corresponds to selecting a locally defined Jeffreys prior. Note that the mode of the likelihood function, the *m.l.e.* is invariant to transformation. If $\widehat{\theta}$ is the *m.l.e.* in the original parameter and we rescale the likelihood using $\phi = g(\theta)$, then the *m.l.e.* in the new parameterization is given by $\widehat{\phi} = g(\widehat{\theta})$.

Given the large number of models in ecology and biology having nonlinear form, good practical advice regarding parameterization and the choice of prior is to:

1. Choose an appropriate likelihood function
2. Parameterize the likelihood function $\phi = g(\theta)$ for stability and interpretability.
3. Using this parameter scale, choose a non-informative prior or an informative prior for ϕ based on existing knowledge and literature.

5.2.9   Model Mis-Specification: Nonlinearity

As nonlinear patterns may underlie many theoretical and real-world relationships among measured variables (see Brimacombe (2016)), the use of a linear model may lead to model mis-specification. In this setting the fitting of a linear model using frquentist or Bayesian approaches may not capture the true underlying relationships among the variables in the data. This effect may also arise in linear models using inappropriate scaling of specific variables or when clustering effects in the data collection process are disregarded.

The mis-specification effect can be examined generally to some extent. To see this write the linear model as function of two sets of variables:

$$y = X\beta + \varepsilon = X_1\beta_1 + X_2\beta_2 + \varepsilon$$

where $n > p$. Assume the variables of interest are grouped in the X_1 ($n \times p_1$) matrix with p_1 variables, the X_2 ($n \times p_2$) matrix has p_2 additional variables, where $p_1 < p_2$ and $p_1 + p_2 = p$. The error term ε ($n \times 1$) is assumed to have $\varepsilon \sim N(0, \sigma^2 I)$. The main goal here is to estimate a model for the variables in the X_1 matrix.

Now assume there is a true nonlinear model underlying the X_1 set of variables that describes the true theoretical relationship among the X_1 variables. Re-express the linear model to represent this fact:

$$y = F(X_1\beta_1) + X_2\beta_2 + \varepsilon$$

where $F(X_1\beta_1)$ is a known nonlinear model for the X_1 subset of variables. Let's use this model to develop an estimation approach. Replace $F(X_1\beta_1)$ with its Taylor expansion about β_{10} to the first derivative to give:

$$y = [X_1\beta_{10} + F'(X_1\beta_1)(\beta_1 - \beta_{10})] + X_2\beta_2 + \varepsilon. \tag{5.7}$$

So if a linear model is applied in this setting we can interpret this as using a local linear approximation rather than the true nonlinear model, giving:

$$y = X_1\beta_{10} + X_2\beta_2 + \varepsilon^*$$

where $\varepsilon^* = \varepsilon + F'(X_1\beta_1)(\beta_1 - \beta_{10})$. Fitting this linear model we will (i) potentially miss the nonlinear aspect of the data and (ii) apply an approach which employs a biased error distribution since we now have $\varepsilon^* \sim N(F'(X_1\beta_1)(\beta_1 - \beta_{10}), \sigma^2 I)$.

In light of these potential effects it is often important to understand the basic structures in the data before assuming basic theoretical structures. See also Brimacombe (2016).

5.2.10   Introduction to Basic Survival Analysis

The application of survival analysis has become common in the investigation of time to response data in relation to many types of infections and diseases in humans and animal species. Standard analysis of time response data in these settings often requires the data being transformed to achieve approximate normality or homogeneous error to better justify application of standard ANOVA methods, generalized linear models, or more typically nonparametric ANOVA rank-based testing procedures.

In longer time frames, responses tend to become correlated within subject due to multiple observations on individual animals and survival time becomes an important response subject to more complex assumptions requiring more involved statistical models. The use of survival analysis in this type of setting rather than standard ANOVA methods is often motivated by an attempt to more carefully understand the distribution of infection times and to formally compare the hazard rate (risk) of infection across, for example, exposure levels. The idea of proportional hazards among the treatment group arises here as well as the use of modified likelihood based analysis. Collett (1994) is a standard basic reference.

It is sometimes useful to view only some of the parameters in a model as random. This use of random effects has a long history in animal experiments. The extension of random effects to survival analysis is termed frailty modelling. See for example Aahlen (1994). Practically, this refers to settings where some animals may be immune to the particular hypothesized exposure-disease setting. These animals are less frail than others and are usually randomly distributed throughout the experiment. If a marker can be found that correlates with this suspected immunity, such as a marker gene, this variable can be given a random effects interpretation in the model.

In aquatic species, for example, where careful breeding and environmental practices are often incorporated, frailty can be seen as a general measure of the genetic and environmental similarity of the animals. If the experimental controls are adequate, survival models adjusted for frailty should fit no better than usual survival models.

Survival analysis permits the use of regression concepts, but in settings where use of the normal error distribution is often inappropriate. The use of regression approaches with a different set of underlying distributions appropriate for time response data include Weibull regression and exponential regression models. These can be seen as components of generalized linear models. In these settings the basic fitting principles are not based on least squares, but rather maximum likelihood or Bayesian estimation.

Survival analysis techniques offer experimental researchers a useful set of tools specifically developed for time response data. Concepts such as hazard rate (risk of infection or onset of the event of interest at a specific time, given survival up to that time) and survival curves (taken from the life table methods of actuarial science) along with the assumption of proportional hazards, give researchers a broader set of analytical tools.

5.2.10.1 Survival Analysis Modeling

In settings where the response of interest is time to infection or survival time once infected, a set of statistical models and approaches have been developed that focus on the risk of onset as the central aspect in modelling time related response variables. This risk or "hazard" is often formally modelled using distributions such as the exponential distribution (which reflects an assumed constant risk of infection) and the Weibull distribution (which reflects a potentially increasing or decreasing risk of infection). Once these are examined and the best fitting one chosen, overall differences in time to onset across the treatment groups can be assessed using maximum likelihood based approaches. The Bayesian perspective can also be applied here with suitable prior densities. Nonparametric approaches with a focus on hazard and time based responses are also available.

In many species exposure time is often positively correlated with risk of infection or hazard rate. The degree of risk may be further related to type of species, temperature level, intensity of water flow, type of feeding and other variables. In such cases a regression type survival data model may be used to relate time as a response to these variables or control for their potential influence on response time. The Cox proportional hazards model is the most flexible regression model appropriate for this purpose, as are more formal regression models based on the exponential and Weibull distribution. The Cox model is more flexible due to its semi-parametric nature, but achieves this by assessing risk in relation to a user-chosen baseline group. The model fitting technique is usually a variant of large sample based maximum likelihood estimation.

The notion of censored data also plays a role in the application of survival methods. Some animals, over the course of the experiment, may not become infected. For these animals, infection time may lie beyond the endpoint of the experiment and will not be observed. This often occurs when new treatments are being tested. In a sense the end of the experiment censures observation of the actual infection time. To adjust for this, the probability of the unobserved infection time lying at or beyond the endpoint of the experiment is included in any calculations regarding risk. This is done through adjusting the overall likelihood function.

In some experimental settings it may be the case that there are no censored data points. In this case censoring based adjustment to the likelihood function is not appropriate. It is also worth noting that censored data points (infection time beyond the end of the experiment) and immunes (probability zero of infection) will appear in the observed data similarly. How to formally include and interpret the infection times of non-responsive subjects in the analysis is often a matter of interpretation. As many survival concepts have been developed in the observational and uncontrolled data settings of epidemiology, careful thought should be given when applying these concepts. See, for example, Hosgood and Scholl (2001).

Before formal testing or the fitting of regression models, some graphical depiction of survival times may be desired. The most common approach to estimating and displaying survival times is use of the Kaplan-Meier estimator. This is a very general nonparametric approach to graphically expressing the proportion of a sample that survives at any given time during the experiment.

If the actual survival time function for a given subject is given by:

$$S(t) = P(T > t) = 1 - F(t)$$

defined at times $t_1, ..., t_k$ where $F(t)$ is the cumulative distribution for infection times, the Kaplan-Meier estimate of the survival function is given by:

$$\widehat{S}(t) = \prod_{j:t_j < t} \frac{n_j - d_j}{n_j} \qquad (5.8)$$

where $t_1, ..., t_k$ are the set of distinct failure times observed over the length of the experiment or study, d_j is the number of deaths at time t_j and n_j is the number of individuals alive and uncensored just prior to time t_j. This can be plotted as a function of time and compared for each treatment group. If these are parallel there is some evidence for the proportional hazards assumption.

Once this function is defined, it may be used for example to estimate the times at which 50% ($t(50)$) of the sample is infected at a given treatment level. Such approaches are standard in dose-response studies where the percent survival is viewed as a function of dosage. But this type of information is also available for more general experimental settings and may be very useful. In the case of fish farming for example, median survival time may have a direct economic interpretation for industry. Emergency culls may be far more likely to occur if a given percentile level of infection is attained within a specific time period. Indeed infections occurring quickly at early stages of the life cycle could prove highly relevant to the economic viability of a fish farm.

Formal definitions can be given to the above mentioned concepts. The hazard rate (risk of infection at any point in time, given non-infection up to that point) can be defined in terms of the survivor function

$$h(t) = \frac{f(t)}{F(t)} = -\frac{d}{dt}\{\log(S(t))\}. \qquad (5.9)$$

For the exponential distribution we have $f(t) = \lambda e^{-\lambda t}$ and $S(t) = e^{-\lambda t}$.

Its hazard rate can be shown to λ, a constant. This implies an assumption of constant risk or hazard if we assume the exponential distribution. While unlikely, this may sometimes be the case for clinical trials of limited duration.

For an assumed Weibull distribution the density for infection time has the form: $f(t) = \lambda\gamma t^{\gamma-1}\exp(-\lambda t^{\gamma})$ and its survivor function is given by $S(t) = \exp(-\lambda t^{\gamma})$. It follows that its hazard function is given by $h(t) = \lambda\gamma t^{\gamma-1}$. If $\gamma > 1$, the hazard or risk of infection is increasing over time and if $\gamma < 1$, it is decreasing. Note that if $\gamma = 1$ the hazard is equal to λ, showing the exponential distribution to be a specific case of the Weibull distribution.

Maximum likelihood estimation is employed to calculate estimates for the unknown parameters. Once the unknown parameters are estimated, the potential distributions and related hazard functions are fit to the data and the best fitting distribution chosen. If the choices are between Weibull and exponential distributions, this is formally accomplished by statistically testing to see if $\gamma = 1$. Note that in well-understood settings where clinical expectations imply, for example, lower temperatures having clear physiologic consequences on parasite and host metabolism, the Weibull distribution might be applied in spite of lesser accuracy of fit.

Also available are empirical log-cumulative plots for assessing the overall fit of the assumed distribution. For the case of a Weibull distribution, as $S(t) = \exp(-\lambda t^{\gamma})$, if we take logs, multiply by -1 and take logs again this gives:

$$\log\{-\log(S(t))\} = \log\lambda + \gamma\log t. \tag{5.10}$$

Replacing $S(t)$ by the Kaplan-Meier estimate $\widehat{S}(t)$ defined above and (λ, γ) by their estimates $(\widehat{\lambda}, \widehat{\gamma})$, plot $\log t$ versus $\log\{-\log(S(t))\}$. A straight line is the result if the distribution of infection times follows a Weibull distribution. This diagnostic plot is available in most packages supporting survival analysis. Once the Weibull or other distribution is empirically supported, the corresponding hazard function is taken as given when assessing comparisons across treatment groups.

The median survival time $(t(50))$ is the time t such that $S(t) = 0.5$. For a Weibull distribution this is given by:

$$t(50) = \{\frac{1}{\lambda}\log 2\}^{\frac{1}{\gamma}} \tag{5.11}$$

and setting $\gamma = 1$ gives $t(50) = 1/\lambda \log 2$ for an exponential distribution.

To test differences across treatments using survival or hazard functions a regression format is often employed. These survival data regression models are commonly available in most standard statistical packages. The outputs in these settings are not usually based on ANOVA tables (these correspond to maximum likelihood estimation when normality can be assumed). Rather overall measures of fit based on the likelihood ratio test statistic having a large sample chi-square distribution are reported as well as individual tests of significance for each variable based on simple large sample normality.

If, for example, a Weibull distribution is supported by the data or is chosen, then Weibull regression may be performed by regressing infection time, expressed on a relative hazard scale, against treatment variables and covariates. Here differences in average infection times are not the focus. Rather differences in relative risk of infection expressed using the relative hazard scale are statistically assessed and reported.

5.2.10.2   Linear Models in Survival Analysis

Standard intuition regarding the application of regression in clinical settings can be extended to survival analysis. The difference lies in the use of non-normal distributions such as the Weibull and Exponential distributions. The least squares fitting procedure is replaced with a more general maximum likelihood or Bayesian fitting procedure.

Some standard regression formats are the Weibull regression model and the Cox proportional hazards model. These allow for use of covariates in the fitting and and interpretation of main results. As the response is time the additional issue of time varying covariates may arise in some settings.

If the risk of infection for subjects is thought to be increasing or decreasing through time, a parametric Weibull distribution is often appropriate and Weibull regression is often a standard approach to regression and ANOVA type assessments.

Using the notation developed above, the Weibull distribution possesses a property often desired in time related response models, namely that of proportional hazards. It can be shown that if the Weibull distribution applies and we have for example two groups of subjects yielding infection times, then the hazard for one group is simply proportional to the hazard rate of the other. This simplifies the basic mathematics and allows for use of the usual intuition one would apply to regression or ANOVA models, but the scale remains that of hazard. If a formal model is not supportable, then the Cox proportional hazard is a partially non-parametric alternative that can be used here for regression modelling. Most standard statistical packages give results for both Cox and Weibull models.

Using the notation given above, if we assume a Weibull distribution, it implies proportional hazards. To see this we write:

$$h_i(t) = \exp(\beta_1 x_{i1} + \beta_2 x_{i2} + \cdots + \beta_p x_{ip})h_0(t)$$

as the hazard rate for the ith animal where this risk is assumed to be related to covariates $x_1, \ldots, x_p$. For the Weibull distribution the baseline (setting all $x_i = 0$) hazard rate is given by $h_0(t) = \lambda \gamma t^{\gamma - 1}$. Plugging this in and using the result that the density is given by $f(t) = S(t)h(t)$, we can determine the density function modelling the infection time for the ith animal. From this the likelihood function is available and maximum likelihood estimates for all parameters can usually be obtained. Again, this is now a standard feature of most common statistical packages.

Using the maximum likelihood estimated values $\widehat{\beta}_1, \widehat{\beta}_2, \ldots, \widehat{\beta}_p, \widehat{\lambda}, \widehat{\gamma}$, the estimated survival function for the ith individual is given by:

$$\widehat{S}_i(t) = \exp[\exp(\widehat{\beta}_1 x_{i1} + \widehat{\beta}_2 x_{i2} + \cdots + \widehat{\beta}_p x_{ip})\widehat{\lambda}\widehat{\gamma}t^{\widehat{\gamma}-1}].$$

Similarly for given values of $x_1, \ldots, x_p$ (typically age of animal, temperature of water, specific stressors) the hazard function given above can be estimated giving $\widehat{h}_i(t)$, and plotted as a function of time. These are the typical graphical outputs of regression using survival analysis. As linearity, even on a hazard scale, cannot simply be assumed, plots of $\log(\widehat{S}_i(t))$ versus the log of the cumulative hazard function can be examined for linearity. Residuals are also available for plotting for general assessment of the assumed Weibull model.

Also available is the estimated value for the median survival time defined above:

$$\widehat{t}(50) = \{\frac{1}{\lambda}\log 2\}^{\frac{1}{\widehat{\gamma}}}$$

The actual fitting of models can be carried out in a fashion analogous to the fitting of standard regression models. Terms are dropped or added from the hazard function (rather than the linear regression function in typical applications) based not on partial-F statistics, but rather changes in the likelihood statistic, and the asymptotic chi-square distribution of -2log $\widehat{L}$ is used as means of determining best fitting models. Typically even stepwise regression procedures are available.

5.2.10.3   Random Effects in Survival Settings

The concept of frailty involves attempting to assess how stable the hazard rate and associated survival model and analysis are when potential random effects parameters are present in the design of the experiment. For example, immune animals may be present in the two groups. This will affect any significant differences observed through the well-known regression to the mean effect. At worst it may lead to severe bias and invalidate the results of the experiment. This is discussed in Aahlen (1994).

If a specific probability of an animal being immune can be hypothesized then the hazard functions listed above can be altered to reflect some degree of immunes. This is done by adding a parameter to the hazard functions and also giving the parameter a distribution, similar to the use of random effects in standard least squares and ANOVA based models. Usually the output gives a set of possible survival data results corresponding to a set of potential proportions of immunes in the treatment groups. Frailty is a specific type of robustness assessment of the overall results and can be computed using standard statistical packages.

Example

A basic example is given where right censored survival time data for two treatment groups is examined using both nonparametric Kaplan-Meier estimates

TABLE 5.6
Basic Survival Data Analysis

Treatment 1											
Survival Time	9	14	16	22	27	18	20	15	13	19	25	10
Censored	0	1	1	1	0	1	0	0	1	1	0	1

Treatment 2											
Survival Time	13	14	5	3	4	22	6	12	11	7	21	8
Censored	1	1	1	1	1	0	1	0	1	1	0	1

and Weibull based parametric analysis. Censored data points have Censored = 1. The data are given in Table 5.2. The mean survival time for treatment group 1 is 17.3($\pm$1.6) and treatment group 2 is 10.5($\pm$1.8). The respective median survival times are 16 and 8.

The assumption of model structure should be made carefully. The data is initially examined using the Kaplan-Meier approach for each treatment group separately. This shows no clear distinction between the groups (Figure 5.1). The log rank test can be used to test for homogeneity across groups. This is not significant (p-value = 0.63).

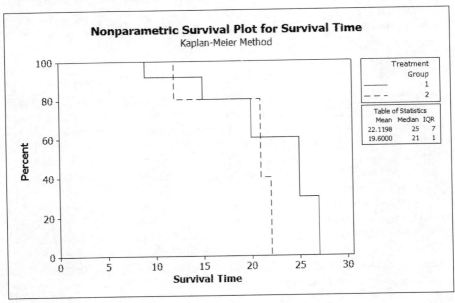

FIGURE 5.4
Kaplan-Meier Survival Plot

To use the Weibull distribution here we initially fit the data to the Weibull distribution for each group. Figure 5.2 shows a probability plot supporting the assumption.

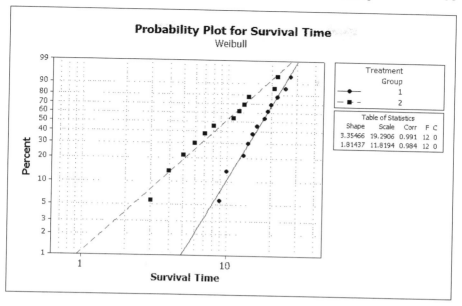

FIGURE 5.5

Probability Plot for Survival Data

The survival curves are now fit using the survival function for the Weibull distribution in Figure 5.3. The graphic shows proportional hazards and the smooth parametric curves of a parametric distribution. The treatment difference is not clearly significant. The smooth nature of assumed parametric distributions especially in smaller sample sizes should be considered carefully.

5.2.10.4   Comparisons to Standard Methods

The use of standard nonparametric ANOVA analysis for this type of data remains commonplace. This approach often takes average response time as the statistic of interest and average differences across exposure groups at specific time points as the basis for comparison. As this data is non-normal, rank based methods are often used to assess significance. This may be acceptable in some clinical settings.

The use of survival based techniques alters the scale of comparison from average time to response to estimated risk or hazard of infection. It also allows for estimation of percentiles related to time of onset and for expression of the number of infected occurring as a proportion related to risk. It allows for the use of distributions such as the exponential and Weibull that better reflect the underlying nature of the data and allow for the inclusion of assumptions regarding increasing or constant risk, a very practical reality in

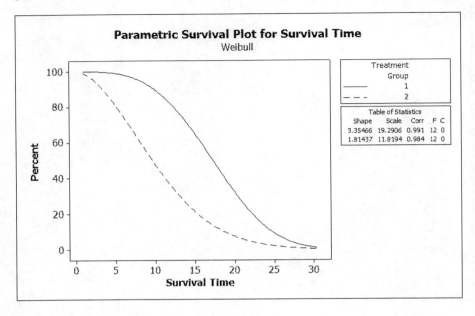

FIGURE 5.6
Weibull Plot for Survival Data

many species related experimentation. Formal modelling of censoring is also possible if required.

In terms of practical implementation, the two approaches give rise to the following suggested strategies for analysis:

A. ANOVA

1. Design the experiment.
2. Determine the scale of response. Rescale the data values if necessary to achieve normality or stable variation.
3. Perform ANOVA based testing, using transformed and/or ranked data values.
4. Adjust ANOVA using covariates or regression concepts if desired.
5. Assess residuals and assumptions made regarding variation.

B. Survival Analysis

1. Design the experiment.
2. Determine possible distributions for survival times.
3. Fit and compare these distributions. Graph log-log plot. Choose best fitting or most appropriate for study goals.

4. Plot survival times (Kaplan-Meier estimator or parametric distribution based). Examine for proportional hazards if using Cox regression.

5. Formally test for differences in overall risk across group (specific test or regression format). Use large sample likelihood based testing or Bayesian perspective.

6. Assess additional parameters such as $t(50)$.

7. Adjust for covariate effects using regression concepts (based on Weibull, gamma or exponential distributions). Use large sample likelihood based testing or Bayesian perspective.

8. Examine residual plots and goodness of fit assessment of chosen hazard model.

The basics of survival analysis have been presented here. The area is very detailed. See Collett (1994) for more detailed development.

5.3 Estimation and Testing

The frequentist theory of statistical inference is often based on the derivation of pivotal quantities. These are functions of both statistics and parameters $Q(T(X), \theta)$ with a known distribution that can be used for testing and confidence intervals. While this could have been somewhat overwhelming with a different statistic and underlying geometry for each modelling situation, R.A. Fisher was able to discern a general principle (Fisher, 1922) that guides the development of most pivotal quantities for many practical situations. This was the likelihood function or likelihood surface in higher dimensions.

In the context of the likelihood function there are three commonly used pivotal quantities available for estimation and testing. Typically they are used in relation to larger samples, though may be used for small samples in many situations. They are also most appropriate in terms of testing when comparing hypotheses that yield nested models; models where the null hypothesis leads to a restricted sub-model of the initial overall model.

Given the likelihood function $L(\theta; y)$, the maximum likelihood estimate (*m.l.e.*) for θ is the value of θ that maximizes the likelihood function, ie.

$$\widehat{\theta} = \max_{\theta}\{\Omega : L(\theta; y)\} \tag{5.12}$$

This is an optimal estimate for θ in the sense that the statistics underlying the likelihood function are typically complete and sufficient, and if such statistics exist, the *m.l.e.* is often a function of these statistics. This is most easily attained if the assumed density underlying the likelihood function is drawn from the exponential family of distributions.

Another important quantity is the Fisher information, evaluated at the *m.l.e.*, either its expected version (I) or observed version (J):

$$I(\theta) = -\left[E \frac{\partial^2 \ln L(\theta; y)}{\partial \theta^2} \right] \tag{5.13}$$

$$J(\theta) = -\frac{\partial^2 \ln L(\theta; y)}{\partial \theta^2} \tag{5.14}$$

This statistic is important as its inverse gives the accuracy of the *m.l.e.* estimate and provides an optimal information bound, the Cramer-Rao bound, $I(\widehat{\theta})^{-1}$, for the accuracy that can be obtained by the *m.l.e.* under standard regularity conditions. Geometrically it represents the expected or observed local curvature in the log-likelihood about $\widehat{\theta}$. To estimate θ, the standard large sample 95% confidence interval is given by:

$$\widehat{\theta} \pm 1.96 \cdot I(\widehat{\theta})^{-1/2}$$

Note that a more general perspective is to view the Fisher information $I(\theta)$ as a measure of (negative) entropy for the entire model-data combination viewed as a single integrated system. Entropy speaks to the level of predictability present in the overall model-data combination; a general measure of goodness of fit. The applicability of the Fisher information to many complicated and often nonlinear systems and models gives it an importance in a wide variety of scientific, engineering and environmental applications. See, for example, Frieden (2004).

The three large sample pivotal quantities typically derived from the likelihood function are:

1. **Likelihood ratio statistic:** This compares the (log) likelihood value at the overall or unrestricted mode $\widehat{\theta}$, which is viewed as reflecting the model under the alternative or non-restricted hypothesis, to the (log) likelihood value evaluated at the restricted mode $\widehat{\theta}_0$, reflecting the model under the null hypothesis. The degrees of freedom q for the large sample chi-square sampling distribution is given by the difference between the number of free parameters under the null and the number of free parameters for the unrestricted or alternative hypothesis:

$$L_1 = -2 \ln[L(\widehat{\theta}_0)/L(\widehat{\theta})] \sim \chi_q^2 \tag{5.15}$$

and large values of L_1 imply rejection of the null hypothesis $H_0 : \theta = \theta_0$.

2. **Wald statistic:** This is based on a direct comparison of the mode of the likelihood function, the *m.l.e.* $\hat{\theta}$ with the hypothesized null value θ_0:

$$L_2 = \frac{(\hat{\theta} - \theta_0)}{\sqrt{1/I(\hat{\theta})}} \sim N(0,1) \qquad (5.16)$$

and large values of L_2 imply rejection of the null hypothesis $H_0 : \theta = \theta_0$.

3. **Score Statistic:** This is based on the first derivative of the log-likelihood function. It can be shown that under the null hypothesis, the expected value of the score is zero, and we have as a test statistic:

$$L_3 = S(\theta_0)/\sqrt{I(\theta_0)} \sim N(0,1) \qquad (5.17)$$

and large values of L_3 imply rejection of the null hypothesis $H_0 : \theta = \theta_0$.

In standard frequentist analysis, these basic pivotal quantities are placed in the context of repeated sampling and p-values from which an inference regarding $H_0 : \theta = \theta_0$ versus the respective alternative $H_a : \theta \neq \theta_0$ can be drawn. Note that these three pivotal quantities should agree in terms of significance when the sample size is large. Cox and Hinkley (1974) and Casella and Berger (2001) are standard references. These results are especially useful in generalized linear models where they are usually the basis of frequentist inference.

Likelihood functions can be poorly behaved. There may be multiple maxima or simply a highly unstable shape to the likelihood leading to poorly behaved test statistics. Examples of this include the likelihood function related to use of the logistic regression model in small samples (Hauck and Donner, 1977), the 3-parameter Weibull model used in reliability and survival analysis (Smith and Naylor, 1987) and settings where the number of parameters is linked to the number of observations; the well known Neyman-Scott problem. In some settings, re-parameterization can help stabilize the likelihood and reduce these difficulties (Cook and Tsai, 1990; Cox and Reid, 1987). Likelihood based inference in some ecological models can inherit underlying chaotic properties. See Givens and Poole (2002).

5.3.1 Assessing Significance

Generally speaking, to assess a fitted or observed value for a statistic or set of statistics from a frequentist perspective, we can employ the ideas of sampling theory or the related ideas of re-sampling theory, ie. the computer based bootstrap resampling method and its variants. Sampling theory derives primarily from the ideas of probability theory and in particular the central limit theorem in relation to parameter estimation. We can also use the calculus of probabilities and obtain small sample exact distributions of some statistics in specific situations. If testing hypotheses, these calculations should be conducted assuming the null hypothesis is true. If they are a component of power calculations, then they can be derived under broader assumptions.

On a practical level significance often reflects the idea of probability itself; a long-run proportion that exists under many theoretical replications. In applied settings this can be viewed as being supported by randomization. Typically this is summarized by a frequency distribution or histogram. We imagine a long series of independent replications of the experiment, each giving a value of the statistic of interest. These values and their rates of occurrence are then summarized by the "sampling distribution" of the statistic in question. Technically this is slightly altered to reflect a cumulative sequence of the statistic, and we consider the frequency behavior of the statistic, typically a mean, or standardized mean. The laws of large numbers, the Chebyshev inequality and central limit theorems apply in this setting. We do not attempt a detailed review of probability here. In Part IV these concepts are employed as appropriate.

A $100(1 - \alpha)\%$ confidence interval for a population mean μ, for example, is usually determined pre-experimentally using the sampling distribution of an estimator for μ:

$$P(\overline{x} - c_{\alpha/2} \cdot \frac{s}{\sqrt{n}}, \overline{x} + c_{1-\alpha/2} \cdot \frac{s}{\sqrt{n}}) = 1 - \alpha$$

where c_α is the relevant percentile of the sampling distribution of $\overline{x}$. The probability coverage given by the interval is technically $1 - \alpha$. The accuracy of this depends on whether the probability calculation is an exact result or approximate. Some coverages, especially when using approximation of the underlying model in some nonlinear regression models, can be inaccurate.

Note that the $1 - \alpha$ aspect is technically pre-experimental. Once the data is collected and values for the statistics are entered, the probability the interval contains the parameter value is actually 0 or 1. This detail tends to be left aside in most frequentist usage of confidence intervals.

A Probability Calculus

In small samples the theoretical joint distribution of the sample is the starting place and by methods of calculus, changing variables and marginalizing via integration, exact distributions can be derived. In practical settings with a likelihood function available, the *m.l.e* is typically the statistic of interest and if an exact distribution is not available, then large sample normality results can be derived under regularity conditions.

There is a structure to sampling theory results. A simple calculus underlies large sample results and exact results where we can assume normally distributed data. Further, many standard distributions fall into two types of distributions; location-scale and exponential family distributions.

If we, for example, assume an *i.i.d.*sample $z_1, ..., z_n$ from a $N(0,1)$ distribution, we have the following as a basic template:

$$\sum_{i=1}^{n} z_i^2 \sim \chi_n^2$$

$$\frac{z_i}{\sqrt{\frac{\sum_{i=1}^{n} z_i^2}{n}}} \sim t_n$$

$$\frac{\sum_{i=1}^{m} z_i^2 / m}{\sum_{i=m+1}^{n} z_i^2 / (n - m)} \sim F_{m, n-m}$$

and these distributions are used often in regards to deriving p-values and confidence intervals.

If a likelihood is not available, or not available to sufficient detail, so-called non-parametric statistics can be examined. These are typically based on sign-ranked values which have exact distributions or large sample based results. In small samples where we cannot assume normality we may also employ a non-parametric approach to derive sampling distributions of standard statistics. This can be done using the computer based bootstrap resampling method.

5.3.1.1 Generic Bootstrap Procedure

Computer based resampling techniques can also be accessed to estimate the distribution of a statistic of interest, likelihood based or otherwise. The bootstrap resampling procedure (Efron, 1987) can be generically defined:

1. Collect the original sample data $y_1, ..., y_n$.

2. Generate a new (bootstrap) sample, often of the same sample size, selecting with replacement. This gives $y_1^*, ..., y_n^*$ which are statistically independent in terms of probability. Calculate the statistic of interest $T = T(y_1^*, ..., y_n^*)$.

3. Repeat 2. B times, typically with $1000 < B < 10,000$. This gives the values $T_1, ..., T_B$ for the statistic T.

4. Examine the empirical distribution (histogram) and related statistical properties of T. For example the bootstrap estimate of the population characteristic average is given by taking the average of the $T_1, ..., T_B$ values:

$$\widehat{T}_{Boot} = \frac{1}{B} \sum_{j=1}^{B} T_j$$

5. The bootstrap distribution may require correction for bias. The BC_a method (Efron, 1987) can be applied here.

The accuracy of the $\widehat{T}_{Boot}$ estimator can be estimated by its standard error:

$$S_B^2 = \frac{1}{B-1} \sum_{j=1}^{B} (T_j - \widehat{T}_{Boot})^2$$

or a more robust estimator

$$S_B^{MAD} = \frac{1}{B-1} \sum_{j=1}^{B} |T_j - \widehat{T}_{Boot}|$$

When a model is available, depending on the focus of the statistical inference, bootstrap resampling is typically performed on the vector of observed residuals which respects the original fit of the observed data and the structure of the fitted model itself. These residuals are then resampled and used to generate pseudo-observations from which the statistic values are generated.

In a linear model these would be of the generic form:

$$y^* = X\widehat{\beta} + \widehat{\varepsilon}_{Boot}$$

The statistic of interest T is then based on resampled y^*, along with the original X and $\widehat{\beta}$. Thus the bootstrap distribution of any statistic, in such a linear model setting, can be viewed as conditional on the observed residual vector. This includes the bootstrapped least squares estimator, for example, $\widehat{\beta}^* = (X'X)^{-1}X'y^*$. Note further that the entire dataset (y, X) can also be resampled to generate bootstrap values and this may be useful in settings related to regression models.

References and related exercises for this chapter may be found at the end of Chapter 6.

6

Introduction to Bayesian Statistical Methods

6.1 Bayesian Approach to Statistical Modeling and Inference

The Bayesian approach underlies some of the earliest applications of statistical models and probability, for example by Laplace (Stigler, 1986). See also Bernardo and Smith (1994), Chapter 1. However it is the frequentist work of Pearson, Fisher and others in the field of population genetics in the early 1900s that is closer to the beginning of modern statistics. Pearson, with chi-square test for contingency tables, principal components analysis, method of fitting distributions and frequentist, sampling theoretic outlook, had a perspective understandable to modern researchers employing statistical methods.

R.A. Fisher extended the work of Pearson in many directions, further developing the analysis of contingency tables, creating the ANOVA table and related models and analysis as well as modern approaches to regression and correlation, multivariate analysis, combinatorics, extending statistical methods in biometry, genetics and agriculture (Box, 1978). Importantly he defined and developed the likelihood function (Efron, 1998) and related Fisher information. That said, he is often remembered in the history of science as a population geneticist, whose work helped lay the mathematical foundation for Mendelian genetics and population genetics in general including the Fundamental Theorem of Natural Selection. See Edwards (2000).

Harold Jeffreys was a geologist who wrote on the topics of information and probability theory from a Bayesian perspective. His work, Theory of Probability (1939), examined statistical inference from a Bayesian perspective and he suggested choosing a prior density in such a way as to be non-informative where the likelihood was pronounced based on the Fisher information and thus provided one of the central objective rules which guide the selection of prior densities in many settings. This has often been re-visited. See for example, Robert et al., (2009). In recent times, the likelihood-dependent Bayesian approach of Jeffreys and others has been accepted by many researchers, with the understanding of the implicit need for care in such models with nonlinear or hierarchical elements and a respect for the surprising difficulties often encountered when integrating the likelihood-posterior density function in higher dimensions.

A useful early book for a review of basic principles from a Bayesian statistical perspective is Box and Tiao (1973). The overlap with likelihood based frequentist models is apparent and the extensions and clarity that can be provided by the Bayesian context presented and discussed. There is also a focus on robustness of inferences in relation to prior selection. See also Bernardo and Smith (1994) and O'Hagan (1994) for more theoretical approaches to Bayesian statistics.

While application of Bayesian methods were slow to develop due to the computational challenges of multiple integration in higher dimensions (to obtain marginal posteriors for example), Bayesian methods have a long history in the statistical literature, especially from a subjective perspective. This is most strongly expressed in the work of Savage (1962) and Good (1984) with a strong underlying utilitarian argument. As all researchers begin with a subjective perspective (prior density), which is then updated upon viewing the data (through the likelihood and resulting posterior density) science can be viewed as a sequential updating of information, an updating that can be formally placed in a Bayesian probabilistic context.

6.1.1   Priors and Posteriors

The formalization of prior beliefs, typically in regard to a specific set of population characteristics, is a formal component of designing an experiment from the Bayesian perspective. But typically in many applied settings, this is approached in a less structured manner. As given in Chapter 1, the likelihood function can be written:

$$L(\theta \,|\, data) = c \cdot \Pi_{i=1}^{n} f(y_i; \theta) \qquad (6.1)$$

with the related joint posterior given by:

$$p(\theta \,|\, data) = k \cdot p(\theta) \cdot L(\theta \,|\, data) \qquad (6.2)$$

where k is the relevant constant of integration and $p(\theta)$ is the prior distribution for θ. This is the primary context for inference from the Bayesian perspective. If θ is a vector then the marginal posteriors for the respective θ_i element of θ is required. Typically the expected value of the posterior density, $E[\theta|y]$, is taken to be the point estimate of θ. The mode is also available.

An interval estimate or credible region for θ is given by (a, b) where:

$$\int_{a}^{b} p(\theta \,|\, data) d\theta = 1 - \alpha$$

where a and b are chosen such that the interval (a, b) is of minimal length. This can be termed a highest posterior density (HPD) interval.

6.1.2   Modeling Strategy

As models in biology and ecology are very often expressed in terms of nonlinear mathematical relationships and related probability models, there are some practical steps that can be followed when looking to apply Bayesian methods in a given model. Once a carefully determined design has been selected for the experiment or observational study in question (randomization, adequate sample size, issues of control and potential causation clearly formulated), there are several steps that can help guide the development of a Bayesian analysis.

The list of issues to address in developing an analysis given in Chapter 1 above can be extended:

- Carefully select the likelihood function. More than one underlying probability density may be applicable to the problem and there more be more than one viable likelihood form to investigate. This may also require examination of the literature.

- Stabilize the parameterization in the model. Typically this may reflect a re-scaling of the parameters in the model giving a more linear scale or orthogonal Fisher information variance-covariancce matrix. Or it may need no alteration.

- Choose a prior or set of priors for the selected (re-scaled) parameters. Justify these selections.

- Be careful to set the prior before examining the current dataset. Prior beliefs should be formally defined before beginning any analysis.

- Obtain the joint posterior density and fitted overall model.

- Obtain the marginal posteriors, posterior odds ratios and Bayes factors as required for inference.

- Examine the stability of the joint posterior and resulting marginal posteriors, posterior odds ratios, Bayes factors and resulting inferences for the set of priors.

In Part IV we apply this approach to a set of applications and examine results both independently and with reference to frequentist-likelihood results.

6.1.3   Some Standard Prior Choices

The most easily supported type of prior distribution is the objective or non-informative prior. It is surprising how flexible the definition of non-informative can be. Factors influencing how a non-informative prior can be defined include the range of the parameters in question, the context within which the non-informative aspect is to be defined (likelihood function, Fisher Information, overall model, measures of entropy) and the degree of non-linearity present in the model (implying potential re-scaling of the parameter space and how this affects the non-informative aspect).

As discussed earlier, there are several choices available for prior distributions and here we discuss the most common. Note that all should result in a proper joint posterior density to be useful:

1. Subjective priors.

2. Non-informative (often improper) priors. An example is the uniform distribution.

3. Jeffreys prior (likelihood and information based), which is locally non-informative.

4. Conjugate priors which provide interpretability and easier calculation.

5. Matching priors. These match the coverage probabilities of credible intervals to their frequentist versions.

6. Bernardo reference prior reflecting an information theoretic approach.

7. Convenience prior. The rise of advanced numerical integration techniques based on Monte Carlo probabilistic methods has lead to settings where the prior to be selected aids in the convergence of the computing technique.

6.1.4   Example: Normal Sample

Consider again a sample of n individual measurements y_i, collected independently from a common population having a probability distribution following the normal distribution with mean μ and unknown standard deviation σ. We write this as $y_i \sim N(\mu, \sigma^2)$.

A typical prior density that can be employed is of the form:

$$p(\mu, \sigma) = c \cdot \frac{1}{\sigma}$$

where c is a constant. To see why, we change variables $\sigma \to \log \sigma$. This implies:

$$p(\mu, \log \sigma) = c$$

which is a non-informative (flat prior) for the parameters $(\mu, \log \sigma)$. As the log transformation is $1 - 1$ this is acceptable and typical of the manipulations often used to determine acceptable priors for a specific model, on some scale.

The posterior density for (μ, σ) is then proportional to the multiple of the prior and likelihood:

$$p(\mu, \sigma | y) = c \cdot \sigma^{-1} \cdot \sigma^{-n} \cdot \exp\left[-\frac{1}{2\sigma^2} \sum (y_i - \mu)^2\right].$$

As the interest lies typically with the marginal distributions for μ and σ, we need to integrate out the unwanted parameter. If the unwanted parameter can be independently estimated or assumed known, conditional posterior distributions are also available, but these are less frequently employed as the methodology of integration has improved.

The marginal distribution for μ here can be written:

$$p(\mu|y) = \int p(\mu, \sigma|y)d\sigma.$$

Writing $\sum(y_i - \mu)^2 = [(n-1)s^2 + n(\bar{y} - \mu)^2]$ and using the gamma integral gives the Student-t distribution with $n - 1$ degrees of freedom. We can write this result as:

$$\frac{\mu - \bar{y}}{s/\sqrt{n}}|y \sim t_{n-1}$$

and the resulting ± 2 standard deviation central regions are identical to the usual 95% frequentist confidence interval for μ. In many settings, especially where there are nuisance parameters to average out, the Bayesian result will be similar to the frequentist, but typically with narrower intervals. This may be viewed in terms of shrinkage (see below for some discussion) and reflects the number of nuisance parameters, some of which may originate in the prior.

While MCMC based integration has widened the set of operational priors and posterior densities available for analyses, it is useful to note several standard densities that arise in the Normal sample case from a Bayesian perspective.

6.1.5 Example: Linear Model

We can write the standard linear model with independent and identical errors as:

$$y_i = \beta_0 + x_{1i}\beta_1 + x_{2i}\beta_2 + \cdots + x_{pi}\beta_p + \varepsilon_i = X\beta + \varepsilon$$
$$\varepsilon_i \sim N(0, \sigma^2)$$

where $i = 1, ..., n$, $X = (x_1, x_2, ..., x_p)$ an array of column vectors, one for each explanatory variable, $\beta' = (\beta_0, \beta_1, \beta_2, ..., \beta_p)$ and $\varepsilon' = (\varepsilon_1, ..., \varepsilon_n)$.

The corresponding likelihood can then be written in matrix notation:

$$L(\beta, \Sigma|data) = c \cdot |\Sigma|^{-1/2} \exp\left[\frac{-1}{2}(y - X\beta)'\Sigma^{-1}(y - X\beta)\right].$$

Prior selection for the set of β_i parameters often assumes they are independent of the variation matrix Σ and each other. Several distributions are then typically employed as prior densities for the β_i parameters. These include the multivariate Student-t density, the multivariate normal, various other symmetric distributions and non-informative (improper) uniform distributions. These options are available for example in the WINBUGS and most R Bayesian software.

For the variance parameter matrix, Σ, it can have several structures. A typical approach assumes all correlations equal to zero and a set of independent variances σ_i^2, though this may be further simplified to homogeneity ($\sigma_i^2 = \sigma^2$ for all i) and is then written $\Sigma = \sigma^2 I$. We can also assume $\Sigma = \sigma^2 C$ where C is a matrix of weighting constants.

The multivariate posterior density for β can be obtained. Assuming $\Sigma = \sigma^2 I$ with σ^2 unknown, a typical prior for (β, σ^2) is given by:

$$p(\beta, \sigma^2) = p(\beta) \cdot p(\sigma^2) = \sigma^{-2}$$

With the assumed normal likelihood, this results in a joint posterior density following the multivariate t-distribution with the standard least squares estimators as key elements:

$$p(\beta | y, X) = \frac{\Gamma(1/2(v+p))|X'X|^{1/2}s^{-p}}{[\Gamma(1/2)]^p \Gamma(v/2)v^{p/2}} \left(1 + \frac{(\beta - \widehat{\beta})'(X'X)(\beta - \widehat{\beta})}{vs^2}\right)^{-1/2(v+p)}$$

where

$$\widehat{\beta} = (X'X)^{-1}X'y$$

$$s^2 = \frac{(y - X\widehat{\beta})'(y - X\widehat{\beta})}{n - p}$$

and $v = (n-p)$. From this, posterior density based inference for each individual β_i is available using the univariate Student-t distribution.

Without existing data or pre-existing studies it follows that several different priors should be assumed and employed when modeling data, with the resulting posterior based inferences examined for robustness and stability across the overall set of priors.

Limitations on Model and Prior Selection

From a technical perspective, not all prior density shapes, when combined with a likelihood function, will yield practically useful or proper posterior distributions. The resulting posterior density may not properly integrate to one or may be unstable for inference. In high dimensional settings where the MCMC approach must be used to achieve the marginal posteriors of interest, lack of convergence may arise as an implication of such problems. While researchers applying the Bayesian approach to likelihood must ensure the resulting posterior density is well behaved, it is difficult, for example, in nested design or hierarchical settings to guarantee this good behavior (Kass and Wasserman, 1996).

From a practical perspective, the selection of priors will often reflect the degree to which the scientific setting accepts or requires additional assumptions beyond those underlying the development of a likelihood function. As we will see in Part IV, there are settings in which the additional assumptions of Bayesian statistics provide useful inferential settings, but with the underlying caveat that the analysis is conditional on the relevance of the additional prior assumptions.

6.1.6 Information Sensitive Priors

If the concept of *posterior information* is defined to be the local curvature of the log-posterior about its mode, we can write:

$$-\frac{\partial^2 \ln p(\theta|data)}{\partial\theta^2} = -\frac{\partial^2}{\partial\theta^2}[\ln c + \ln p(\theta) + \ln L(\theta|data)]$$

$$= -\frac{\partial^2}{\partial\theta^2}\ln p(\theta) + \frac{\partial^2}{\partial\theta^2}\ln L(\theta|data)$$

$$= -\frac{\partial^2}{\partial\theta^2}\ln p(\theta) + J(\theta)$$

where $J(\theta)$ is the observed Fisher information $J(\theta) = -\frac{\partial^2}{\partial\theta^2}\ln L(\theta|data)$ and we focus on choosing a prior that is non-informative, then assuming standard regularity conditions apply to the likelihood function, we can set a second order condition on the prior density:

$$\frac{\partial^2}{\partial\theta^2}\ln p(\theta) = 0$$

or

$$\frac{\partial^2}{\partial\theta^2}\ln p(\theta) = k$$

where k is a constant.

Information similar priors chosen are in this sense non-informative to the second order and are of the general form:

$$\ln p(\theta) = a\theta^2 + b\theta + d$$

where a, b, d are constants.

This gives an exponential family class of prior distributions which may or may not be conjugate to the posterior density. In the large sample setting, up to a multiplicative constant, the observed Fisher Information is the basis of the Bayes posterior information. The normal, binomial and Poisson distributions, for example, satisfy this restriction. The restriction does imply that posterior densities which are third or higher order polynomials in θ are ruled out. These include, for example, $p(\theta) \propto \sin(g(\theta))$ or $p(\theta) \propto \exp(\theta^3)$.

It can be argued that the Jeffreys prior in large samples often achieves this information similar effect. The Bayesian asymptotic result $\theta \sim N(\hat{\theta}, J^{-1}(\theta))$, implies the Jeffreys prior $J(\theta)$ can be interpreted as the inverse of the observed variance $(J^{-1}(\theta))^{-1} = J(\theta)$. As a function this will be relatively flat where the likelihood is pronounced and will be approximately non-informative, preserving the local shape of the likelihood about its mode as a key element in determining the shape of the posterior density.

6.1.7   Example

Consider again the nonlinear regression with Normal error:

$$y_i = \eta(x_i; \beta) + \varepsilon_i$$

where the x are fixed, the ε are *i.i.d.* $N(0, \sigma^2)$ and $\eta(\cdot)$ represents a nonlinear regression surface. Let β be a scalar parameter. The Jeffreys prior is given by:

$$p(\beta) = \frac{F(x; \beta)' F(x; \beta)}{\sigma^2}$$

where $F(\cdot)$ is the first order derivative of $\eta(x_i; \beta)$ with regard to β, and σ is assumed known. This will be acceptable in terms of information similarity if it has the property:

$$\frac{\partial^2}{\partial \beta^2} \left[\ln \frac{F(x; \beta)' F(x; \beta)}{\sigma^2} \right] = k$$

This implies that the nonlinear regression surface should not be too complex as a function of β if it is to be useful as a basis of such likelihood dependent prior selection. See Brimacombe (2016) for more discussion.

6.2   Bayesian Estimation: Marginal Posteriors

The greatest area of advance in applied Bayesian methods has been the development of posterior sampling based methods to carry out multiple integration and obtain marginal posteriors (MCMC and one of its versions, Gibbs sampling). The sampling in question is carried out in the parameter space Ω. While the methods were not developed *per se* for Bayesian application, they have driven much of the computational advances that have allowed for practical application of Bayesian methods.

If interest lies in a specific parameter, say θ_1, the marginal posterior for θ_1 is generally defined as:

$$p(\theta_1 | y) = \int \cdots \int p(\theta_1, \theta_2 \ldots \theta_p | y) \, \partial\theta_2 \partial\theta_3 \cdots \partial\theta_p \tag{6.3}$$

WINBUGS, software used here, employs Markov Chain Monte Carlo (MCMC) and Gibbs sampling to carry out the required marginalizations. See for example, Gilks et al., 1996. Many packages in R use similar methods.

There are caveats regarding the application of MCMC, typically related to the shape of the joint posterior density which is to be sampled and integrated. Convergence is not guaranteed and diagnostics may be required to assess the convergence of the method. Indeed it is surprising how difficult it can be to integrate higher dimensional likelihood/posterior surfaces. Details can be found in Gilks et al., (1996) and Robert and Casella (2010). This is also further discussed in the context of the applied examples in Part IV.

6.2.1 Normal Approximation

Other approximate methods to obtain marginal posteriors exist and are typically available for theoretical and large sample settings. The normal approximation for the posterior distribution, sometimes referred to as the Bayesian Central Limit theorem, is related to a version of the frequentist central limit theorem.

Assuming a large, independent sample from a common distribution from which we derive the likelihood function $L(\theta \mid data)$ and assuming a prior $p(\theta)$, often non-informative or flat, the resulting posterior density $p(\theta \mid data)$ can be approximated by a normal distribution. We can write this result:

$$\theta \sim N(\widehat{\theta}_{mle}, J(\theta)^{-1})$$

where J is the observed Fisher information and $\widehat{\theta}_{mle}$ the maximum likelihood estimate. Note that J can be replaced by the (expected) Fisher information I.

The accuracy of this approach improves as the underlying posterior density is closer in shape to a normal distribution. The convergence of the posterior to the normal distribution also allows for application of properties of the Normal distribution, including determination of all marginal posterior densities for θ_i and various linear combinations of θ_i.

6.2.2 Laplace Approximation

The Laplace approximation is another approach to estimation based on higher order Taylor expansions. While developed for estimating functions of the underlying posterior, such as a mean or mode, it can be used to obtain marginal posteriors. If we want to estimate the marginal posterior for θ_1 we can initially write:

$$p(\theta_1 \mid data) = c \cdot \int \cdots \int \exp\{\ln p(\theta_1, \theta_2, \theta_3, \ldots, \theta_p \mid data)\} \partial\theta_2 \cdots \partial\theta_p.$$

Assuming the posterior has a unique mode, we can expand the exponent in the integral about the modal value using an appropriate Taylor expansion. Simplifying and using the definition of the maximum value gives an improved approximation to the marginal posterior or functions of it such as a mean value. The details are usually referred to as Laplace's method. It is not unique to statistical application.

The estimate can be written:

$$\widehat{p}(\theta_1 | \, data) = c' \, |V(\theta_1)|^{1/2} \exp\{\ln p(\theta_1, \widehat{\theta}_2(\theta_1), \widehat{\theta}_3(\theta_1), \ldots, \widehat{\theta}_p(\theta_1)| \, data)\}$$

where $(\widehat{\theta}_2(\theta_1), \widehat{\theta}_3(\theta_1), \ldots, \widehat{\theta}_p(\theta_1))$ is the maximum of $-\ln p(\theta_1, \cdot | \, data)$, c' is a normalizing constant and $V(\theta_1)$ the inverse of the second derivative matrix (Hessian) of $-\ln p(\theta_1, \cdots | \, data)$, evaluated at $(\widehat{\theta}_2(\theta_1), \widehat{\theta}_3(\theta_1), \ldots, \widehat{\theta}_p(\theta_1))$. This density upon simplification is a function of θ_1 alone and can be used to derive 95% credible regions for estimation or inference. Further references regarding this approximation can be found in Tierney and Kadane (1986).

This method is better used for measures of central location. It is less accurate as an estimate of the entire marginal posterior density, especially as the number of parameters increase, and works best where the initial joint posterior density is unimodal. When it is not unimodal, the accuracy of the estimate is questionable in the tails of the marginal distribution.

6.2.3   Monte Carlo Probability Based Estimation

In this book we will typically use the MCMC or Gibbs sampling approach underlying the WinBugs program to provide estimates of marginal posterior densities and related estimates. This is based on the Monte Carlo sampling approach that provides a more accurate estimate for the entire marginal density. This approach views an integral of interest as the expectation of an indicator function for the region of interest. This allows for estimation of integrals using central limit theorem and law of large numbers probability based estimates that converge to the integral.

This is not a book on numerical integration, as such we do not present material on the many variations of the MCMC approach in higher dimensions. Versions of this underlie the application of Bayesian methods in Part IV. Several accessible references are given in the Suggested Readings section below. Note that underneath the details, there is a simple goal: integrating out parameters from the joint posterior that are not of interest. This method can also be employed in frequentist settings where a marginal distribution is required, for example in applications of conditional inference. See Robert and Casella (2010).

6.3 Testing: Measures of Evidence

The frequentist and Bayesian approaches differ in approach and interpretation for the testing of specific parameter value hypotheses. From the Bayesian perspective, once the parameter space Ω has been weighted using the posterior, the question arises as how to use and express this information. Posterior odds ratios and Bayes factors are the standard choices and the goal in developing such measures of evidence are stability and interpretability. There is no need to assume a "true" value under which a sampling theory is developed. Rather, evidence from the model and data for the various parameter values of interest (including the null hypothesis value) can be directly compared in terms of posterior density function values. One's beliefs, from the baseline of the prior density, can then be updated and expressed using the observed posterior. The frequentist approach assumes a null value for θ is true and looks for evidence against it, finding evidence for alternatives via contradiction of the null value.

6.3.1 Posterior Odds Ratios

Given the posterior density $p(\theta_1, \theta_2 \ldots \theta_p|\, y)$, the most intuitive approach to comparing its values, which are probabilities lying in $[0, 1]$, is to use odds ratios, which have a long history of application and interpretation. Here we look at the odds ratios of posterior probabilities for the various parameter values being compared. In particular, the posterior odds ratio comparing a specified value θ_0 to a general alternative θ can be written

$$\frac{p(\theta_0|data)}{1 - p(\theta_0|data)} \bigg/ \frac{p(\theta|data)}{1 - p(\theta|data)}$$

and on a log scale this is a difference given by

$$\log\left(\frac{p(\theta_0|data)}{1 - p(\theta_0|data)}\right) - \log\left(\frac{p(\theta|data)}{1 - p(\theta|data)}\right).$$

Typically, we would take the posterior mode or median as the general value against which to compare the null hypothesis. The interpretation given to odds ratios in general is subjective, but typically an odds ratio of at least 1.5 to 2.0 is considered meaningful.

6.3.2 Bayes Factors

Another approach to using probabilities on Ω to examine hypotheses, is to view hypotheses as defining models and examining the change in probability values as we go from the initial overall model to the null model of interest or potential best fitting model. If we start for example with the prior distribution $p(\theta_1, \theta_2 \ldots \theta_p)$, which reflects our beliefs regarding the θ_i values, without any input from the current model-data combination, and want to assess the contribution of the model and data to our probability assessment for the values of θ_i, we might examine the Bayes factor:

$$BF = \frac{p(\theta_1, \theta_2 \ldots \theta_p \mid y)}{p(\theta_1, \theta_2 \ldots \theta_p)} \qquad (6.4)$$

which is essentially a ratio of posterior probability incorporating likelihood information to the initial prior probability incorporating no likelihood based information. On a log scale this is simply the difference between the two, and measures directly how the probability or level of belief for $\boldsymbol{\theta}$ changed as we incorporated the likelihood model-data based information into our assessment

$$\log BF = \log p(\theta_1, \theta_2 \ldots \theta_p \mid y) - \log p(\theta_1, \theta_2 \ldots \theta_p)$$

The representation in 6.4 corresponds to a global assessment of the set of variables comprising the model.

We can also more importantly compare nested models, for example:

$$BF = \frac{p(\theta_1, \theta_2 \ldots \theta_p \mid y)}{p(\theta_1, \theta_2 \ldots \theta_{p-1} \mid y)}$$

to assess the relative impact of not including the X_p variable (setting $\theta_p = 0$) in a linear model setting. In a sense this corresponds to examining the degree to which the hypothesis $\theta_p = 0$ is relevant to posterior probability assessment. We can continue in this manner, defining the various levels of interest corresponding to various sub-models and the underlying "hypotheses" they reflect versus the initial overall model. We keep the variables whose deletion from the model lead to large changes. Note however that we need to conceptually define new prior densities for each nested model. This is questionable. If the prior is a component of the model we should not be modifying key components in the model-data combination as we move through the results of a sequence of tests.

The Bayes factor can be used to compare the relative fit of models. Given two models, we calculate the marginal distribution of the observed data under each of the models respectively. The Bayes factor is then given by:

$$B = \frac{p(y \mid Model1)}{p(y \mid Model2)}$$

and the model with the higher probability is supported by the data. Typically we interpret a Bayes factor in relation to selecting Model 1 over Model 2 in terms of:

Bayes Factor Value	<1	1-3	3-20	20-150	>150
Interpretation	negative	weak	positive	strong	very strong

See Efron et al., (2001). Often this is re-interpreted on a positive/negative/ indeterminate scale. There is no guarantee that a finding of indeterminate can be avoided.

Some caveats regarding the existence of the Bayes factor should be noted. Bayes factors may become infinite in some settings. This may arise when improper priors are used. They may also exist when they should not. See for example Raftery (1996) and Aitkin (1998). This may arise both generally and when specific model structures are employed, such as hierarchical models. Also, whenever the posterior is difficult to obtain, being unstable under integration, typically the Bayes factor will also be problematic. In such settings the posterior odds ratio is to be preferred.

Again, this approach is of course very different from frequentist hypothesis testing where sampling theoretic calculations assume the null hypothesis true and we look for evidence against the null. The Bayesian approach is a learning model approach that allows for assessment of the degree to which specific hypotheses are relatively supported by the observed model-data combination.

6.3.3   Example: Linear Model

Following O'Hagan (1994), the overall fit of competing linear models can be examined using the Bayes factor. If we assume a standard linear model of the form:

$$y = X\beta + \varepsilon$$
$$\varepsilon \sim N(0, \sigma^2 I)$$

where β is a p-dimensional parameter vector and take as the prior for (β, σ^2) the conjugate Normal-Inverse-Gamma distribution:

$$p(\beta, \sigma^2) = Ng(a, d, m, V) = c \cdot (\sigma^2)^{-(d+p+2)/2}$$
$$\times \exp(-[(\beta - m)'V^{-1}(\beta - m) + a]/(2\sigma^2))$$

where a, d, m, V are hyper-parameters typically set at some reasonable value, the resulting posterior is the multivariate Student-t distribution.

If we want to examine the comparative fit of an alternative model, typically reflecting a different choice of explanatory variables but with similar structure, we can use a Bayes factor. Take as the alternative model;

$$y = W\beta_W + \varepsilon$$
$$\varepsilon \sim N(0, \sigma^2 I)$$
$$p(\beta_W, \sigma^2) = Ng(a, d, m_W, V_W)$$

and we obtain the distribution of the response variable y under each model.

The Bayes factor to compare these models is given by:

$$B = \frac{f_W(y)}{f(y)} = \frac{|V|^{1/2}|V_W^*|^{1/2}}{|V_W|^{1/2}|V^*|^{1/2}} \left(\frac{a^*}{a_W^*}\right)^{d^*/2}$$

where $d^* = d + n$, $V^* = (V^{-1} + X'X)^{-1}$, $V_W^* = (V_W^{-1} + W'W)^{-1}$, $m^* = (V^{-1} + X'X)^{-1}(V^{-1}m + X'y)$, $m_W^* = (V_W^{-1} + W'W)^{-1}(V_W^{-1}m_W + W'y)$, $a^* = a + m'V^{-1}m + y'y - m^{*\prime}(V^*)^{-1}m^*$, $a_W^* = a_W + m_W'V_W^{-1}m_W + y'y - m_W^{*\prime}(V_W^*)^{-1}m_W^*$.

The actual interpretation of the resulting Bayes factor is straightforward. The value of B is sensitive to the comparative levels of variation in the model. The Bayes factor will favor the model having the lower relative level of overall residual variation (a definition of better fit).

6.3.4   Model Selection Criteria

While the testing of specific parameter values and the population characteristics they represent is often of interest, as important is the determination of best fitting models. In the frequentist-likelihood setting this is typically carried out by comparing a series of nested models using a series of likelihood ratios or partial F-tests. At each step, if the change in the likelihood ratio is sufficient, we move to the more parsimonious model. We stop when the changes in the likelihood are no longer significant. This approach often tends to fall under the general stepwise model fitting concept. For example, if the original likelihood for the full model is given by $L(\theta_1, \ldots, \theta_p|y)$ and the null hypothesis is, for example $H_0 : \theta_1 = \theta_2 = \theta_3 = 0$, then the frequentist test statistic is given by the likelihood ratio test which has a large sample chi-square distribution (with $q = 3$ degrees of freedom for this hypothesis).

In the Bayesian setting, models and the hypotheses that define them are somewhat interchangeable and are examined using probabilities defined directly on the parameter space. While Bayes factors and posterior odds ratios are standard, there is also a need to assess relative model fit that is in some sense independent of prior densities. Several measures of the information provided by a model (and thus its usefulness) have been developed. They are often extensions of the likelihood ratio statistic.

Typically we are comparing a single full model (M1) which is then restricted to give a second, nested model (M2). Often the restriction will set several parameters $= 0$ and we are interested in whether this new, smaller model (M2), loses an "acceptable" amount of information to be viewed as a useful, parsimonious summary of the overall model. Typically we will have a listing of the various possible models (corresponding to sets of explanatory variables in a linear model for example) each with a given information value. These will then be interpreted relative to the information value in the initial overall model. We take the most parsimonious model that still provides a useful fit.

From a frequentist perspective, the change in the likelihood ratio statistic $(-2\ln LR)$ is of interest in these settings and can be assessed for significance by comparison to a chi-square value with $p_1 - p_2$ degrees of freedom (the number of parameters set equal to zero). If it is not significant, the null hypothesis is not rejected and the parameters of interest can be set equal to zero and the simpler M2 version of the model selected.

The Bayes Information Criteria (BIC) for the models M1 and M2 is defined as:

$$BIC = -2\ln LR - (p_1 - p_2)\log n$$

Similarly, the Akaike Information Criteria (AIC) is given by:

$$AIC = -2\ln LR - 2(p_1 - p_2)$$

Both reflect the difference in the log-likelihood corresponding to moving from M1 to M2, penalized by the overall number of parameters (which can be seen as artificially increasing the accuracy of the models).

The values observed for BIC and AIC in a sequence of nested models should be interpreted carefully, on percent change basis. Note that the BIC measure is related to the Bayes factor (BF) corresponding to the likelihoods underlying M1 and M2 respectively and in large samples we can write:

$$\exp(-\frac{1}{2}BIC) \approx BF$$

Note also that while we have used M1 (full model) as a reference, these measures can also be used to compare various restricted models in relation to each other as well. See Bernardo and Smith (1994) for more details.

6.3.5 Model Averaging

In the context of applying the Bayesian approach to derive a best fitting or optimal model, software is available to average the set of potential models that are generated, rather than looking only at the best fitting model. Often the difference between sets of potential models using the BIC or AIC criteria can be difficult to interpret (and relatively very small). As well, the best fitting models may not all use the same scaling in regard to the data. For example some may have examined survival time (continuous) rather than survival (yes/no).

A type of model averaging can be carried out by averaging a common effect size Δ over the set of posterior densities related to the various models being considered. This can be written:

$$p(\Delta|data) = \sum_j p(\Delta|M_j, data)p(M_j|data)$$

where M_j are the set of models under consideration. See for example Hoeting et al. (1999) for details.

6.3.6   Predictive Probability

The Bayesian context allows for an interesting tool for the examination of model fit and assessment; the predictive distribution. It typically involves developing the initial posterior density of interest and modifying it to determine the probability of an additional independent future observation, y_{n+1} then averaging out all the underlying parameters β_i in the model.

It can be written:

$$p(y_{n+1}|\,data) = \int p(y_{n+1}, \beta_1, ..., \beta_p|\,data)\,d\beta_1 \cdots d\beta_p \qquad (6.5)$$

This probability can be used to predict future values of y, assess the quality of the underlying model, and can also be re-calculated for each element in a sequence of underlying models and thus used as a measure of goodness of fit. See, for example, Meng (1994) for the development of a related tail area interpretation or predictive p-value defined on the sample space. Note that the need for a large number of integrations (all the elements of β must be integrated out) to achieve the predictive density has often limited the use of the predictive density in applied studies.

As an example of its use, consider the linear model described above. In this setting, the predictive density for a future observation is similar in form to the posterior density; the univariate or multivariate Student-t distribution depending on the dimension of the predictive value y_{n+1}. For example, in a simpler setting where inference for a normal sample is considered, the predictive density for a single future value of y is given by the univariate t-distribution(mean, standard error, degrees of freedom):

$$St\!\left(\bar{y}, \left(1 + \frac{1}{n}\right)^{1/2} s, n - 1\right)$$

where $\bar{y}$ and s are the sample mean and standard deviation of the initial sample. In most applied settings, there is no closed form for the predictive density (nor the posterior for that matter) and the MCMC method of approximate integration must be applied. See also O'Hagan (1994).

6.4   Hierarchical Structures and Modeling Approaches

Many sampling designs reflect a naturally occurring hierarchy in regard to how the data is collected. For example to assess the relationship between a specific chemical fertilizer and crop output we may need to collect data on several crops located within several comparable fields. Or to assess a new pharmaceutical treatment used to treat a rare disease, we will need to determine treatment effects within cohorts of patients drawn from several clinics. Such designs must be approached carefully as not all parameters will be relevant to all components of the design.

There is also the potential for a naturally occurring hierarchy within the mathematical formulation of Bayesian methods, which may or may not reflect hierarchy in the sampling design. This occurs when the definition of the prior density for a parameter θ itself depends on other parameters ϕ (which must also be given prior densities). This may reflect the way in which researchers have processed existing beliefs and knowledge rather than the sampling design itself or may simply be a mathematical convenience. The hierarchical setting is therefore more prevalent in the Bayesian formulation and actually allows for a simple approach to defining Bayesian models in complicated design settings. See Lindley and Smith (1972). An example of the hierarchical approach allows us to see how it affects the definition of likelihood.

Consider a study where genes for the onset of a rare disease are identified using a sample of individuals drawn from several different countries. We record yes/no for each gene as the response of interest. For any joint posterior density we can use the definition of conditional probability to re-express the joint prior probability, written here as a function of two vector parameters:

$$f(\beta_1, \beta_2 \mid y) = p_1(\beta_1) * p_2(\beta_2/\beta_1) * L(y \mid \beta_1, \beta_2) \qquad (6.6)$$

with gene effect (β_2) nested within country (β_1) and $p_2(\beta_2/\beta_1)$ representing the conditional distribution of gene within country. Writing out the likelihood and prior components of this model, including the use of a log odds (logit) rescaling of the response y, we have:

$$y_i \sim Bernoulli(p_i) = p_i^{y_i}(1 - p_i)^{1-y_i}$$
$$logit(p_i) = \sum_j X_{ij}\beta_j \qquad (6.7)$$
$$\beta \sim Normal(0, B) \qquad (6.8)$$

The hierarchical effects are modeled in this setting through the prior covariance B which is a diagonal matrix having 0 for each corresponding variance of the β_1 components (gene effects) and differing variances σ_j for each element of β_2 (country effects). This results in country specific effects being given the priors:

$$\beta_{2j} \sim N(0, \sigma_j^2), \, j = 1, \ldots, K.$$

The joint posterior density for the β_{1j} parameters is of interest here, and the central regions for these will reflect the variation in the β_{2j} parameters. Marginal posterior densities for each β_{1j} can be obtained using WINBUGS for example and central inference intervals (mid-95% posterior regions) determined.

In general, care must be taken with hierarchical models. Not all likelihood and prior combinations can be assumed to generate stable posteriors. See for example Kass and Wasserman (1996).

6.5 Empirical Bayesian Approach

In settings where we can accept using previously observed data to aid in the determination of prior densities, an "empirical" Bayesian approach is possible. Typically this implies a two-stage procedure where the observed data is first used to estimate secondary parameters in the assumed prior density. This empirically adjusted prior is then used to derive the posterior density and marginal posteriors of interest, modes and central 95% regions, for the key parameter(s) of interest.

As this is a multi-stage procedure, it has a natural application in hierarchical or nested design settings and can be applied very widely. Typically when the model of interest can be expressed in hierarchical form, the empirical Bayes technique can be used to justify estimating secondary parameters in some manner.

The approach is flexible and can be widely applied. Indeed if the underlying distribution for the likelihood is unknown, the "empirical" aspect of the model can be the use of non-parametric estimates for the sample probability density. Carlin and Louis (1996) provides a reference for the various approaches available within the empirical Bayes approach.

The approach is also useful in settings where there is a large number of similarly structured individual summaries for a common parameter or outcome which must then be combined into an overall estimate, with adjustment for multiple comparisons. See for example Efron (2008) in regards to microarray data where each platform or array contains expression data on thousands of individual genetic sites or regions and overall modeling is helpful in providing a context to address multiple comparison issues.

6.5.1 Example: Two-Stage Normal Model

A common approach when little is known regarding the distribution of a response, is to assume normality. A two-stage normal model for an outcome and underlying parameters can be written in the form:

$$y_i \sim N(\delta_i, \sigma^2)$$

which is interpreted as the sampling distribution for each (independent) y_i given a mean value of δ_i. We take

$$\delta_i \sim N(\theta, \omega^2)$$

as the prior for δ_i given θ (an underlying overall mean) with variance ω^2. Typically σ^2 and ω^2 are assumed known.

Using the fact that the joint distribution can be obtained by multiplying the appropriate conditional and marginal densities and integrating out δ_i, we can obtain the marginal density for the outcome y_i as a function of θ. With some algebra (and the properties of normal distributions) it can be shown that each y_i has an independent $N(\theta, \sigma^2 + \omega^2)$ distribution. This can then be used to estimate the secondary parameter θ. Call this empirical Bayes estimator $\widehat{\theta}$.

An estimated posterior for each δ_i can then be obtained:

$$p(\delta_i | data, \widehat{\theta}) = N(m\widehat{\theta} + (1-m)y_i, (1-m)\sigma^2)$$

where

$$m = \frac{\sigma^2}{\sigma^2 + \omega^2}$$

$$\widehat{\theta} = \overline{y} = \frac{1}{n}\sum y_i$$

Inference is then based on central 95% Bayesian credible regions, using $p(\delta_i | data, \widehat{\theta})$ as the relevant marginal posteriors.

In earlier days, before advances in computing, much focus was placed on deriving the formula for the modal estimate $\widehat{\delta}_i$. These calculations are not practically interesting given the availability of WINBUGS and other software programs and the focus on central regions for inference rather than point estimators, but they do show an interesting aspect of empirical bayes as an approach to estimation. In the two-stage Normal model case considered here, the formula for the modal estimate of δ_i is given by:

$$\widehat{\delta}_i = m\overline{y} + (1-m)y_i = \overline{y} + (1-m)(y_i - \overline{y}).$$

Examination of this provides an insight into the behavior of estimates obtained using empirical Bayes methods. They often can be seen as a weighted average, blending together information from the various sampling sites or subjects in a weighted and intuitive manner. This is due to empirical Bayes being related to the James-Stein approach to estimation, an approach which has both frequentist and Bayesian justification (Efron and Morris, 1977; Efron 2010).

Here, the ith site or subject is the source of all information when $m = 0$, and the entire sample contributes equally through the overall average when $m = 1$. When $0 \le m \le 1$, the estimate blends individual subject or site information with information from the entire sample.

It is important to note the need to ensure that the different samples involved are truly compatible. It is rare that information regarding location or scale is useful when blended together inappropriately. Mixing apples and oranges may not be useful. However using overall information from similarly structured sampling contexts to better assess variation or control for sample-wide issues such as multiple comparison corrections may be justified.

6.6 High Dimensional Models and Related Statistical Models

In recent years the development of large databases has lead to the need for high-dimensional statistical models, often with the number of variables far greater than the number of subjects ($p > n$). This implies that the standard linear model is non-identifiable and the least squares estimator does not exist uniquely. These databases typically arise in genomic applications but are also common in marketing, internet analysis, brain imaging, health outcomes research and other areas using technologies generating vast amount of data. Often simple linear models are applied, but with $p > n$ and an associated sparsity restriction which limits the number of model coefficients and related variables that can be significant:

$$y = X\beta + \varepsilon$$

$$\sum_{i=1}^{p} |\beta|^m < t$$

The sparsity restriction can be made more complex, but the basic idea is similar. The assumption of $p > n$ with the sparsity restriction on the linear model has geometric implications for the model-data combination (Brimacombe, 2014). Frequentist approaches to the sparse linear model focus on estimation through forward stage-wise algorithms such as the lars-LASSO approach (Efron et al., 2004).

The Bayesian approach presents a related but different perspective on the model-data combination and is useful in these non-identifiable situations. The actual values of the data variables are conditioned upon and the probability, through application of Bayes theorem and the assumption of a prior density $p(\beta)$ for β, shifts probabilities to the parameter space. Therefore the relatively small n does not technically limit the obtaining of marginal posteriors for each of the overall p variables. In this sense the model is identifiable primarily due to the assumed prior density.

The sparseness restriction can be directly incorporated into the Bayesian perspective. Park and Casella (2008) suggest the following hierarchical description of the Bayesian LASSO model:

$$y|\mu, X, \beta, \sigma^2 \sim N_n(\mu + X\beta, \sigma^2 I)$$
$$\beta|\sigma^2, \tau_1^2, ..., \tau_p^2 \sim N_p(0, \sigma^2 D), D = diag(\tau_1^2, ..., \tau_p^2)$$
$$\sigma^2, \tau_1^2, ..., \tau_p^2 \sim \pi(\sigma^2)d\sigma^2 \prod_{j=1}^{p} \frac{\lambda^2}{2} e^{-\lambda^2 \tau_j^2} d\tau_j^2$$
$$\pi(\sigma^2) = 1/\sigma^2$$

where $\sigma^2, \tau_1^2, ..., \tau_p^2 > 0$. Note that here the model requires further technical limitations for stability.

Thus the dimensional restrictions of the frequentist approach do not limit the Bayesian analysis, but require the assumption of large relative amounts of prior information. That said, Bayesian approaches in many sparse settings have lead to similar answers to those given by the LASSO and its variants. In the case of normal error the log-likelihood aspect simplifies to the least squares criteria and the log-prior provides a sparseness restriction, a very similar problem to the frequentist setting, viewed as a function of the parameters. This remains an area of research interest.

6.7 Summary

This chapter reviews the theoretical tools and concepts available for researchers working to understand real-world data using a likelihood function. From a frequentist perspective, when a likelihood function is available, the maximum likelihood parameter estimates, along with their sampling distributions, are the basis of inference. From the Bayesian perspective, priors are carefully assumed and modulate the likelihood function, giving rise to posterior distributions for the parameters of interest. These reflect personal beliefs, modified by observing the data, regarding potential values for the parameters of interest.

Through the concept of empirical Bayes, the selection of priors can be made more data oriented. This also occurs if a likelihood based prior, for example the Jeffreys prior, is employed. The additional concepts of central Bayes regions, Bayes factors, predictive densities and posterior odds ratios are also available for the examination of hypotheses, relevant sources of information and statistical inference. These tools, from both frequentist and Bayesian perspectives, are further examined and applied in Part IV.

6.8   Applying the Theory

The concepts and approaches examined in Parts I–III are the key approaches that can be applied in the scientific use of modern statistical inference where a parametric model is available. There are many extensions of these ideas in various particular settings, and non-parametric approaches are available (for example the bootstrap), but the essential models, ideas and tools are have been briefly outlined.

In Part IV, the basic foundations of design, data analysis and statistical inference, reviewed in Parts I–III, are applied to a selection of problems drawn from biology and ecology. These problems have been chosen to illustrate the flexibility of statistical methods in dealing with scientific questions as well as to illustrate, compare and practically integrate frequentist and Bayesian perspectives to applied statistical inference and modeling.

In each setting a brief detailed development of the likelihood and Bayesian setting for particular models (for example generalized linear models) is developed. In each example, the following outline is employed to guide the analysis;

1. Scientific overview of problem
2. Data and graphical analysis and investigation
3. Mathematical and statistical models (likelihood)
4. Hypotheses to be examined and parameters to be estimated
5. Statistical inference and results

While this listing is numbered, the actual identification and examination of key aspects of a model and data combination is often iterative, with the application of more than one model being possible, depending on the specific biological or ecological setting. This flexibility is necessary as statistics is an area in which mathematical and statistical models and principles are applied directly to real-world data, in the context of an existing scientific literature. This is always somewhat eclectic by definition and can be surprisingly challenging in practice.

The statistical models to be employed should be justified by data, drawn from the literature or in hand, and robust enough to withstand peer-review criticism. While certain models are obvious in application (for example linear models), the areas of ecology and biology provide a more challenging set of problems, where nonlinear and hierarchical mathematical models, along with complex statistical sampling issues, allow for a more interesting and flexible set of statistical models for application. The statistical approaches both influence and are influenced by the science. Which is, of course, how it should be.

A brief review of the Bayesian approach as representing a distinct perspective on probability and modeling along with additional comparisons to the frequentist approach are given at the beginning of Part IV.

6.9   Bibliography

[1] Aahlen O. (1994). Effects of Frailty in Survival Analysis, *Statistical Methods in Medical Research* 3, p. 227–243.

[2] Aeschbacher S., Beaumont M.A. and Futschik (2012). A Novel Approach for Choosing Summary Statistics in Approximate Bayesian Computation. *Genetics*, 192, p. 1027–1047.

[3] Aitkin M. (1998). Simpson's Paradox and the Bayes Factor. *J. Roy. Statist. Soc. B*, 60, p. 269–270.

[4] Bernardo J.M. and Smith A.F.M. (1994). *Bayesian Theory*. John Wiley and Sons Inc. New York, NY.

[5] Berger J.O. and Wolpert R.L. (1988). *The Likelihood Principle 2nd ed.* Hayward, CA. Institute of Mathematical Statistics.

[6] Box G.E.P. (1983). An Apology for Ecumenism in Statistics, in *Scientific Inference, Data Analysis and Robustness*, Academic Press, Inc., New York.

[7] Box G.E.P and Cox D.R. (1964). An Analysis of Transformations, *J. Roy. Statist. Soc. B* 26, p. 211-252.

[8] Box G.E.P. and Tiao G.C. (1973). *Bayesian Inference in Statistical Analysis*, Addison-Wesley, Reading, Mass.

[9] Box J.F. (1978). *R.A. Fisher: The Life of a Scientist.* John Wiley and Sons Inc.

[10] Brimacombe M. (2014). High Dimensional Databases and Linear Models: A Review. *Open Access Medical Statistics* 2014:4, p. 17–27.

[11] Brimacombe M. (2016). Local Curvature and Centering Effects in Nonlinear Regression Models. *Open Journal of Statistics* 6, p. 76–84.

[12] Brimacombe M. (2016). Some Likelihood Based Properties in Large Samples: Utility and Risk Aversion, Second Order Prior Selection and Posterior Density Stability. *Open Journal of Statistics* 6, p.1037–1048.

[13] Brimacombe M. (2017). High Dimensional Models and Analytics in Large Database Applications. *Actionable Intelligence in Healthcare.* J. Leibowitz, A. Dawson eds., Auerbach Publications.

[14] Brown L. D. (1986). *Fundamentals of Statistical Exponential Families, with Applications in Statistical Decision Theory.* Institute of Mathematical Statistics, Hayward, CA.

[15] Carlin B.P. and Polson N.G. (1991). Inference for Nonconjugate Bayesian Models Using the Gibbs Sampler, *Canad. J. Statist.* 19, p. 399–405.

[16] Carlin B.P. and Louis T.A. (1996). *Bayes and Empirical Bayes Methods for Data Analysis.* Chapman & Hall, New York, NY.

[17] Casella G. and Berger R. L. (2001). *Statistical Inference*, 2nd ed. Duxbury Press, Pacific Grove, CA.

[18] Cook R.D. and Tsai C.-L. (1990). Diagnostics for Assessing the Accuracy of Normal Approximations in Exponential Family Nonlinear Models. *J. Am. Statist. Assoc.* 85, p. 770–777.

[19] Collett D. (1994). *Modelling Survival Data in Medical Research.* Chapman & Hall, New York.

[20] Cox D.R. and Hinkley D.V. (1974). *Theoretical Statistics.* Chapman & Hall, New York.

[21] Cox D.R. and Reid N. (1987). Parameter Orthogonality and Approximate Conditional Inference, *J. Roy. Statist. Soc. B* 49, p. 1–39.

[22] David A. P., Stone M. and Zidek J. V. (1973). Marginalization Paradoxes in Bayesian and Structural Inference. *J. Roy. Statist. Soc. B,* 35, p. 189–233.

[23] DeGroot M.H. (1973). Doing What Comes Naturally: Interpreting a Tail Area as a Posterior Probability or as a Likelihood Ratio. *J. Am. Statist. Assoc.* 68, p. 966–969.

[24] Dobson A.J. (1986). *An Introduction to Statistical Modeling.* Chapman & Hall Ltd., New York, NY.

[25] Draper N. R. and Smith H. (1981). *Applied Regression Analysis, Second Edition,* New York: John Wiley & Sons.

[26] Draper N.R. and Smith H. (1998). *Applied Regression Analysis, 3rd ed.* John Wiley & Sons, Inc. New York.

[27] Eaves D.M. (1983). On Bayesian Nonlinear Regression with an Enzyme Example. *Biometrika* 70, 2, p. 373–379.

[28] Echambadi R., Hess J.D. (2007). Mean-Centering Does Not Alleviate Collinearity Problems in Moderated Multiple Regression Models. *Journal of Marketing Science* 26, 3, p. 438–445.

[29] Edwards A.W.F. (1992). *Likelihood.* The Johns Hopkins University Press. Baltimore, Maryland.

[30] Edwards A.W.F. (2000). The Genetical Theory of Natural Selection. *Genetics* 154, p. 1419–1426.

[31] Efron B. (1998). R.A. Fisher in the 21st Century. *Statist. Sci.* 11, p. 95–122.

[32] Efron B. (2008). Microarrays, Empirical Bayes and the Two-Groups Model. *Statist. Sci.* 23, p. 1–22.

[33] Efron B. (2010). The Future of Indirect Evidence. *Statist. Sci.* 25, p. 145–157.

[34] Efron B. (1987). Better Bootstrap Confidence Intervals (with Discussion). *J. Am. Statist. Assoc.* 82, p. 171–200.

[35] Efron B., Gous A., Kass R.E., Datta G.S., Lahiri P. (2001). Scales of Evidence for Model Selection: Fisher versus Jeffreys. *Lecture Notes-Monograph Series, Vol. 38, Model Selection*, p. 208–225.

[36] Efron B., Hastie T., Johnstone I. and Tibshirani R. (2004). Least Angle Regression. *Annals of Statistics* 32, No. 2, p. 407–451.

[37] Efron B. and Morris C. (1977). Stein's Paradox in Statistics. *Scientific American* 236:119–127.

[38] Fisher R.A. (1922). On the Mathematical Foundations of Theoretical Statistics. *Philosophical Transactions of the Royal Society* A, 222, p. 309–368.

[39] Fisher R.A. (1934). Probability, Likelihood and the Quantity of Information in the Logic of Uncertain Inference, *Proceedings of the Royal Society* A, 146, p. 1–8.

[40] Fisher R.A. (1958). *The Genetical Theory of Natural Selection, 2nd ed.* Dover Publications, New York & Constable Co. Ltd, London.

[41] Fraser D.A.S., Reid N. and Wu J. (1999). A Simple General Formula for Tail Probabilities for Frequentist and Bayesian Inference. *Biometrika* 86, p. 249–264.

[42] Frieden B. R. (2004). *Science from Fisher Information: A Unification.* Cambridge University Press, Cambridge, UK.

[43] Geyer C.J. (2005). *LeCam Made Simple: Asymptotics of Maximum Likelihood Without the LLN or CLT or Sample Size Going to Infinity.* Technical Report No. 643 (revised). School of Statistics, University of Minnesota.

[44] Gilks W.R., Richardson S., Speigelhalter D.J. (1996). *Markov Chain Monte Carlo in Practice.* Chapman & Hall, New York.

[45] Givens G. and Poole D. (2002). Problematic Likelihood Functions from Sensible Population Dynamics Models: A Case Study, *Ecological Modeling* 151, p. 109–124.

[46] Good I.J. (1984). A Bayesian Approach in the Philosophy of Inference. *British Journal for the Philosophy of Science* 35, p. 161–166.

[47] Hauck W.W. and Donner A. (1977). Wald's Test as Applied to Hypotheses in Logit Analysis. *J. Amer. Statist. Assoc.* 72, p. 851–853.

[48] Hicks C.R. (1982). *Fundamental Concepts in the Design of Experiments, 3rd ed.* Holt Rinehart and Winston, New York.

[49] Hodges J.S. (1987). Assessing the Accuracy of Normal Approximations, *J. Amer. Statist. Assoc,* 82, p. 149–154.

[50] Hoeting J.A., Madigan D., Raftery A.F., Volinsky C.T. (1999). Bayesian Model Averaging: A Tutorial. *Statist. Sci.* 14, p. 382–417.

[51] Hosgood G. and Scholl D.T. (2001). The Effects of Different Methods of Accounting for Observations from Euthanized Animals in Survival Analysis, *Prev Vet Med* 48, p. 143–154.

[52] Houggard P. (1982). Parameterizations of Non-Linear Models. *J. R Statist. Soc B,* 44, p. 244–252.

[53] Jeffreys H. (1934). Probability and Scientific Method, *Proceedings of the Royal Society A* 146, p. 9–16.

[54] Jeffreys H. (1939). *Theory of Probability, 1st ed.* The Clarendon Press, Oxford.

[55] Kass R.E. and Wasserman L. (1996). The Selection of Prior Distributions by Formal Rules. *J. Am. Statist. Assoc.* 91, p. 1343–1370.

[56] LeCam L. (1990). On the Standard Asymptotic Confidence Ellipsoids of Wald. *International Statistical Review* 58, 2, p. 129–152.

[57] Lindley D.V. and Smith A.F.M. (1972). Bayes Estimates for the Linear Model. *J. Roy. Statist. Soc. B. 34,* Vol 1, p. 1–41.

[58] McCullagh P. and Nelder J. (1989). *Generalized Linear Models, 2nd ed.* Chapman & Hall, New York.

[59] Meng X.-L. (1994). Posterior Predictive p-values. *Annals of Statistics* 22, p. 1142–1160.

[60] O'Hagan A. (1994). *Kendall's Advanced Theory of Statistics, Volume 2B, Bayesian Inference.* John Wiley and Sons, New York.

[61] Park T. and Casella G. (2008). The Bayesian Lasso, *J. Am. Statist. Assoc.* 103, Vol. 482, p. 681–686.

[62] Pawitan Y. A. (2000). Reminder of the Fallibility of the Wald Statistic: Likelihood Explanation. *The American Statistician* 54, p. 54–56.

[63] Penrose K., Nelson A., and Fisher A. (1985). Generalized Body Composition Prediction Equation for Men Using Simple Measurement Techniques. *Medicine and Science in Sports and Exercise* 17(2), p. 189.

[64] Raftery A.E. (1996). Approximate Bayes Factors and Accounting for Model Uncertainty in Generalized Linear Models. *Biometrika* 83, p. 251–266.

[65] Reid N. (1988). Saddlepoint Methods and Statistical Inference. *Statistical Science* 3, p. 213–238.

[66] Robert C.P. and Casella G. (2010). *Introducing Monte Carlo Methods with R*. Springer, New York.

[67] Robert C.P., Chopin N., Rousseau J. (2009). Harold Jeffreys Theory of Probability Revisited. *Statist. Sci.* 24, p. 141–172.

[68] Savage L. (1962). *Foundations of Statistical Inference: A Discussion*, Methuen, London.

[69] Seber G.A.F., and Wild C.J. (1989). *Nonlinear Regression*. John Wiley, New York.

[70] Smith R.L. and Naylor J.C. (1987). A Comparison of Maximum Likelihood and Bayesian Estimators for the Three-Parameter Weibull Distribution. *Appl. Stat.* 36, p. 358–369.

[71] Sprott D.A. (1980). Normal Likelihoods and Their Relation to Large Sample Theory of Estimation. *Biometrika* 60, p. 457–465.

[72] Stigler S.M. (1986). Laplace's 1774 Memoir on Inverse Probability. *Statist. Sci.* 1, p. 359–378.

[73] Tanner M (1996). *Tools for Statistical Inference: Methods for the Exploration of Posterior Distributions and Likelihood Functions*, 3rd Edition. Springer-Verlag, New York.

[74] Tierney L. and Kadane J.B. (1986). Accurate Approximations for Posterior Moments and Marginal Densities, *J. Am. Statist. Assoc.* 81, p. 82–86.

[75] WINBUGS Software. http://www.mrc-bsu.cam.ac.uk/bugs/winbugs/contents.shtml

6.10 Questions

1. A sample of size $n = 20$ independent biological measurements from a normally distributed sample with known variance σ_0^2 is given by

$y = (5, 4, 6, 4.3, 4.2, 4.4, 6.0, 7.6, 8.1, 6, 5, 6, 4, 5, 5, 6, 4.2, 8.2, 4.5, 4.2)$.
(i) Obtain the likelihood and 95% frequentist confidence interval for the population mean μ. (ii) Develop the Bayesian analysis for conjugate, non-informative (improper) and Jeffreys prior, obtaining for each the central 95% posterior credible region for μ. (iii) What % difference in these credible regions corresponds to the use of the different priors?

2. Assume that $X_1, ..., X_n$ are an $i.i.d.$ sample from the following distribution: $f(x|\theta) = \theta x^{\theta-1}$, for $0 \le x \le 1$ and $\theta > 0$. Find the $m.l.e.$ for θ, and its variance as $n \to \infty$.

3. Examine the case of a simple regression model with centered data so the regression line goes through the origin $(0,0)$. Take as the model $y_i = \beta x_i + \varepsilon_i$, $i = 1, ..., n$. Assume the x_i are fixed constants and the ε_i are random, independent error terms sampled from a $N(0, \sigma^2)$ distribution. (i) Find the $m.l.e.$ for β. (ii) Find its distribution. (iii) Assume a prior density for β of the form $N(\theta, \rho^2)$ where θ and ρ are known. Obtain a 95% credible region for β.

4. In the context of the regression model $y_i = \beta x_i + \varepsilon_i$, $i = 1, ..., n$ where the x_i are fixed constants and the ε_i are independent error terms sampled from a $N(0, \sigma^2)$ distribution, argue that the $m.l.e.$ and least squares estimate for β are identical.

5. Let $X_1, ..., X_n$ be an $i.i.d.$ sample from the Poisson(λ) distribution. (i) Let the prior density for λ be given by a $Gamma(\alpha, \beta)$ distribution. and find the posterior distribution of λ. (ii) Argue that the $Gamma$ distribution is a conjugate prior for the $Poisson$ distribution. (iii) Find the posterior mean and variance of λ. (iv) Show that, from a frequentist perspective, the posterior mean is not an unbiased estimator for λ.

6. Let $X_1, ..., X_n$ be an $i.i.d.$ sample from a $N(\theta, \sigma^2)$ distribution with σ known. Let the prior density for θ be given by $\theta \sim N(0, \sigma^2)$.

   (a) Argue this is a conjugate prior for θ.
   (b) Find the posterior density for θ and find the posterior mean.
   (c) Now let the prior be slightly more resistant to outlier values, $p(\theta) = (1/(2\sigma)) \exp(-|\theta|/\sigma)$ with σ known. Find the posterior density for θ. Find the posterior mean and compare to the posterior mean for the Normal prior.

7. Let $X_1, ..., X_n$ be an $i.i.d.$ sample from the Bernoulli(θ) distribution and let the prior density for θ be given by a $Unif(0, 1)$ density.

   (a) Show that the posterior mean is given by $E[\theta|x] = (\sum_{i=1}^{n} x_i + 1)/(n + 2)$.
   (b) Let $\tau(\theta) = \theta(1 - \theta)$. Show that $E[\tau(\theta)|x] = (\sum_{i=1}^{n} x_i + 1)(n - \sum_{i=1}^{n} x_i + 2)/(n + 2)(n + 3)$.

(c) Show that if $E[\theta|x]$ in a. is viewed from a frequentist perspective it is unbiased for θ. Find its Mean Squared Error (MSE).

8. Let $X_1, ..., X_n$ be an *i.i.d.* sample from a $N(\theta, 1)$ density. Let the prior density for θ be given by $\theta \sim N(\mu_0, 1)$. Show that the posterior density $p(\theta|x)$ is $N((\mu_0 + \sum x_i)/(n+1), 1/(n+1))$.

9. In a large sample size setting:

   (a) The general large sample *m.l.e.* distribution is given by $\widehat{\theta} \sim N(\theta, I^{-1}(\theta))$ where $I(\theta)$ is the Fisher information. For $X_1, ..., X_n$ an *i.i.d.* sample from a Poisson(λ) distribution, derive the large sample distribution for the *m.l.e.* of λ.

   (b) The corresponding large sample Bayesian result is given by $\theta \sim N(\widehat{\theta}, J^{-1}(\theta))$ where $J(\theta)$ is the observed Fisher information. For the Poisson example, compare the Bayesian 95% central credible region to the *m.l.e.* based 95% large sample confidence interval.

10. Let $X_1, ..., X_n$ be an *i.i.d.* sample from the Poisson(λ) distribution. Let the prior density for λ be given by an improper uniform distribution. Compare the resulting posterior density to the posterior density resulting from a prior density for λ given by a $Gamma(\alpha, \beta)$ distribution. Consider the percent overlap of credible regions.

11. Let $X_1, ..., X_n$ be an *i.i.d.* sample from the Poisson(λ) distribution with the prior density for λ given by a $Gamma(\alpha, \beta)$ distribution. Find the posterior distribution of λ and use it to develop a posterior odds ratio appropriate for testing the hypothesis $H_0 : \lambda = \lambda_0$ vs. $\lambda = \lambda_a$. How would we interpret this if the resulting odds ratio was less than 1, equal to 1, greater than 1?

12. For statistical problems in both ecology and biology, the Bayesian approach requires careful and justified selection of the prior density. This typically implies the selection of a non-informative prior. Discuss issues affecting the definition and selection of non-informative prior densities.

13. In a linear model setting we sometimes encounter unexpected anomalies in the data such as heteroscedasticity (non-constant variance). Ecological studies using spatially oriented samples are an example where variation may be a function of distance out from a central reference point in the study area. As well, in biology, consideration of dose-response type models often lead to thresholds beyond which the response in question becomes much more variable. (i) Discuss how heteroscedasticity can be incorporated into the standard linear model formulation by using appropriately reweighted data. This is sometimes called weighted least squares. (ii) Discuss how the ordinary linear model can be extended to incorporate more complicated variance structures, for example by replacing

the assumption of constant variance (homogeneity) σI, with Σ having various patterns of correlation. (iii) Discuss the prior densities that are appropriate for these types of extended models and the form of the resulting Bayesian posteriors.

14. In many biological modeling applications of the linear model, a standard approach to dealing with anomalies in the data is to take a log transformation or the Box-Cox transformation (y^λ) of the response y. This raises issues regarding the interpretation to be given to such rescaling. (i) If we take the Box-Cox transformation of the response y would this alter the definition and stability of the Jeffreys prior (how should we view the λ parameter; as fixed or variable)? (ii) Can data transformations in general affect the selection and stability of prior densities? Should they?

15. The stability of the likelihood function itself and the maximum likelihood estimate can be questioned in standard nonlinear regression model settings. A perspective on this is given by examining the Wald statistic and related confidence regions which can be unstable in nonlinear regression models, especially nonlinear regression with normal errors or logistic regression (generalized linear models). This affects the stability of the inferences drawn from these models. Referring specifically to Pawitan (2000) and Hauck and Donner (1977), discuss these difficulties and how they can affect statistical inference.

16. In ecological settings where there is an underlying growth process or the process develops in a self-referential manner, for example studies of population dynamics, there is growing use of dynamic non-linear models. These have at their core a regression relationship of the form $y_t = f(y_{t-1})$ where $f(\cdot)$ is nonlinear in form. Such models are prone to chaotic behavior and related properties such as fractal (highly nonlinear) 95% contours where elliptical forms are expected. An example of this in a statistical context is the estimation of whale populations. Referring to Givens and Poole (2002) (i) Assuming normal errors, discuss the stability of the likelihood concept in such models. (ii) Discuss the application of the Bayesian approach, how the averaging aspect of the Bayesian approach may smooth and stabilize the inferences obtained. (iii) In many mathematical biological models restrictions or simplifications may be necessary to obtain an answer. List any assumptions made to fit the whale population model.

17. Assume that a researcher is interested in counting the number of infected animals in a herd population using a sample of size n. The binomial distribution, $Bin(n, p)$, often applies well here. As the herd is spread across several farms, a hierarchical sampling design is applied. (i) Assume a binomial distribution describes the probabilities

arising in the sampling process within each subgroup, and derive the respective likelihood function. (ii) Assuming the beta distribution, $Beta(\alpha, \beta)$, as a prior for p, derive the posterior density. (iii) Discuss how you would apply the empirical Bayes approach to derive a marginal model for the underlying parameters α, β. (iv) How would their values be estimated? (v) How would you use these estimated values to provide posterior based inference regarding the parameter of interest p?

18. The number of new cases of an animal disease occurring within a given time frame is often of interest. The Poisson distribution is a useful density to model such occurrences. (i) Obtain the likelihood function when we assume a Poisson distribution, $Pois(\lambda)$, to describe disease occurrences within a given time frame. (ii) Derive the Fisher information I and conditional Fisher information J. (iii) Obtain and discuss the Jeffreys prior for λ.

19. Discuss (i) the use of posterior odds ratios and Bayes factors. (ii) How does this compare and contrast with the more standard scientific setting where a null hypothesis is assumed true and evidence to the contrary is assessed? (iii) Develop an example drawn from ecology or biology where these approaches have or can be applied.

20. From the statistical literature, find and discuss an example where the p-value is shown to be fairly conservative when compared to the Bayesian approach. Discuss how the p-value can be described in purely Bayesian terms. Refer for example to DeGroot (1973).

21. Review the use of random effects. (i) Define the random effects model. (ii) What aspects of the experimental or observational design supports the use of prior distributions in this setting? (iii) Discuss the difference between the fully Bayesian approach to the analysis of a linear model with random effects and the frequentist analysis of the same model. (iv) This approach has often been used in animal breeding in relation to the underlying, unobserved, genetics involved. Refer to the literature and briefly discuss a specific random effects model in genetics, its justification and application.

6.11 Suggested Readings

1. Bernardo J.M. and Smith A.F.M. (1994). *Bayesian Theory*. John Wiley and Sons Inc. New York, NY.

2. Box G.E.P. and Tiao G.C. (1973). *Bayesian Inference in Statistical Analysis*. Addison-Wesley Publishing Company Inc. Reading, Massachusetts.

3. Casella G. and Berger R.L. (2001). *Statistical Inference, 2nd ed.* Duxbury Press, Pacific Grove, CA.

4. Dobson A.J. (1986). *An Introduction to Statistical Modeling.* Chapman & Hall Ltd., New York, NY.

5. Gelman A., Carlin J.B., Stern H.S., Rubin D.B. (1995). *Bayesian Data Analysis.* Chapman & Hall, New York, NY.

6. O'Hagan A. (1994). *Kendall's Advanced Theory of Statistics, Volume 2B, Bayesian Inference.* John Wiley and Sons, New York.

7. Silvey S.D. (1975). *Statistical Inference.* Chapman & Hall, New York, NY.

8. Walker S.G. (2004). Modern Bayesian Asymptotics. *Statist. Sci.* 19, p. 111–117.

Part IV

Applications Using Bayesian and Frequentist Likelihood Methods in Biology and Ecology

7

Case Studies: Bayesian and Frequentist Perspectives

7.1 Preface

Here we briefly discuss practical application of the Bayesian and frequentist approaches as related but distinct perspectives on the application of probability models and the likelihood concept. The frequentist approach has its conceptual roots in the experimentalist view of scientific investigation with direct application to the repeated sampling related structures of population genetics, agriculture biology and time series applications. The modern Bayesian perspective has roots in the areas of insurance, loss and risk, and decision-making, areas that must incorporate risk with limited experimental or sampling data. In these settings, expert belief and experience are key components. From these roots, this perspective has been widely applied.

The frequentist approach avoids formalizing prior knowledge regarding the unknown parameter or population characteristics of interest. In cases where there is a long literature and much known data, this might seem an inefficient use of existing scientific information, to not employ a prior density as a baseline level of knowledge, but the focus is on the assessment of the next sample to be collected, independent of past experience. The underlying stochastic process is modeled via the likelihood function when available and the design of the experiment is randomized. In the Bayesian setting prior belief is updated using model and observed data through a formal prior density and the likelihood function. There is no averaging over all possible sample outcomes and the analysis is conditional on the observed data.

Each perspective has its strengths and weaknesses, though in settings where the underlying probability model can be seen as giving rise to a likelihood function, there is much overlap. There are some that view it necessary that the two approaches agree perhaps through the use of specifically chosen prior densities (matching priors). That perspective is not viewed as essential here. The two, at their core, are distinct and provide, where justified, useful alternative perspectives on the model-data combination. It is possible to do statistics without a prior density. It is also possible to do statistics in settings where repeated sampling is not practical. Note however that in larger samples these two approaches tend to agree, where it makes sense to compare them (the Bayesian structure can be more complex).

In the particular setting of hypothesis testing, the Bayesian approach takes a learning perspective, while the frequentist setting reflects a more conservative experimentalist approach, testing through contradiction. In both approaches, the design of the study, experiment or data collection must be of high quality. It must not be biased or confounded or non-representative and the sample should be drawn at random. This is technically not directly a clear requirement of the Bayesian approach, but a scientific one for all quality studies, however they are to be analyzed.

Recently the advanced use of computational Monte Carlo and Markov Chain Monte Carlo techniques in the context of numerical integration have dominated much application of Bayesian models. This has tended to emphasize the selection of priors that provide the most efficient or convergent marginal posterior calculation.

The weaknesses of frequentist statistics have been discussed many times, with the most obvious being the conservative nature of the approach, especially in terms of p-values, and the lack of clear incorporation of the observed data into most probability calculations. Recent advances in resampling or bootstrap techniques have to some extent attempted to address this issue. Some caveats regarding application of the Bayesian method are given below.

7.1.1   Some Particulars

1. Bayesian probability is not frequentist probability. It is a proxy for personal belief or utility, justified somewhat by de Finetti's work on exchangeable joint distributions and more importantly connections between preferences expressed in terms of probability scales and utility functions. This is a key reason why Bayesian and frequentist results can be placed next to each other as distinct approaches. They reflect different perspectives, overlapping through use of the likelihood function.

2. The selection of a prior density is an important component of the Bayesian design of an experiment. It provides the baseline of available knowledge regarding the parameter of interest. Much current Bayesian application very loosely justifies the choice of prior, if at all. Being able to write down a mathematical formula for a prior does not absolve the researcher of validating its choice as a representation of belief regarding potential values of the parameter in question. The mathematical properties of the prior density should be linked to the personal beliefs/knowledge of the researcher. Determining a prior is a necessary component of the Bayesian design of the experiments and its interpretation.

3. The experimentalist roots of frequentist statistics reflect the "proof

by contradiction" approach of Hume and Popper, a mainstay of scientific thought for several hundred years. The Bayesian approach is a learning model, in which the analysts beliefs are updated by the model-data combination reflected in the likelihood. Today's posterior becomes tomorrows prior. There is no null hypothesis to assume true and prove through contradiction. We simply re-weight the potential values of the parameter of interest as we move through the learning process, collecting evidence in relation to potential values of θ. Note that pivotal quantities are not essential from the Bayesian perspective.

4. The posterior density, the result of the prior likelihood combination, is often a richer function in terms of parameters than the likelihood alone and will typically provide a better fit to the data along with smaller residual error and thus improved statistical efficiency compared with frequentist likelihood methods. This is somewhat misleading and should not be viewed as a justification of Bayesian methods, but rather a secondary smoothing effect to be carefully considered.

5. Placing Bayesian estimators or results in frequentist settings can be done. Bayesian and frequentist estimators can be compared using for example the mean squared error measure or tail areas, but the contexts underlying the estimators themselves differ and can be interpreted separately.

6. When applied in an experimental or observational context, the Bayesian approach conditions on the observed data. By conditioning so completely on the observed data, the Bayesian perspective tacitly assumes the data is of high quality (unbiased, representative, random sample). This may not be a justifiable assumption. Such a conditional approach and its conclusions should be examined through additional replications of the experiment/observation in order to be validated if the data is not randomized. Nature will fool you every time. Note that the frequentist "bootstrap" and randomization/permutation test techniques share the "full conditioning on the observed data" criticism and thus also should be subject to the caveat of replicated data, where possible, in order to be validated.

7. The importance of computational approaches and their requirements such as conveniences priors need to be balanced with the basic structure of the Bayesian approach. The addition of parameters to the model through prior assumption needs to be justified and computational methods should, when possible, be expressed in a way that can be clearly understood, avoiding a black-box approach. The usefulness of priors lie in their reflecting baseline understanding of the science involved.

7.2   Case Study Template

In the case studies that follow each problem considered begins with a brief simplified development of the scientific question at hand and describes the data collection process, the collected data, the statistical model to be applied (likelihood and priors), hypotheses to be investigated and finally the analysis. The analysis reports and interprets both frequentist and Bayesian approaches.

In each setting more complicated analyses are possible. Several additional suggested exercises regarding further analyses, projects and related theoretical results close each section.

The most direct reference for the Bayesian computational aspect is the following;

- Lunn D., Jackson C., Best N., Thomas A., and Spiegelhalter D. (2013). *The BUGS Book: A Practical Introduction to Bayesian Analysis* (Chapman & Hall/CRC Texts in Statistical Science) 1st Edition. CRC Press, Taylor and Francis Group LLC, Boca Raton FL. ISBN-13: 978-1584888499, ISBN-10: 1584888490

8

Case Studies in Ecology

Ecology is one of the more interesting and challenging areas of scientific research. It interfaces and interacts conceptually and in terms of modeling considerations with biology, the basic sciences, environmental science, evolution and other disciplines. Given this diversity, design issues are particularly challenging and must be carefully developed in relation to the particular study or experiment. From this wide set of possible applications, many with specific theories and related mathematical models, we examine several examples that allow us to apply an interesting set of statistical models.

The issues arising in relation to the changing environment have lead to a growing focus on the modeling of ecological phenomena. When the environment alters, the basic parameters affecting many natural populations are modified and may have both direct and indirect effects on various species, including their population density, spatial distribution and basic metabolic components. The relationship between these aspects comprise some of the questions relevant from the ecological perspective. Many factors affect food supply and available natural habitats and these alter evolutionary patterns through time, if not basic survival.

Tracking both the micro and macro level ecological patterns and changes in populations provide ecological (and biological) researchers with a wide variety of research contexts and an equally wide ranging set of study designs and related statistical models. Models in ecology often have a high degree of random effects interpretation often due to challenges in the real-world study designs involved.

In settings such as ecology and biology there often are theoretical expectations that can be expressed in terms of underlying probability and likelihood models and related null values for the key parameters in question. Here we use standard prior choices, but detailed elicitation of prior belief is possible leading to justification of prior density selection and structure from theory.

Here we first discuss the modeling of species abundance in relation to population density and metabolic rates and then examine soil erosion in a northern agricultural environment.

8.1   Biodiversity: Modeling Species Abundance

8.2   Science

The effects of changes in the environment on species is widely and carefully studied in ecology and biology. Whether these changes are caused by the expansion of housing developments in local wetlands or by more global phenomenon such as global warming and related climate change, tracking the abundance of species over time is a challenge. Carefully defined catchment areas, the use of averaged measures across a set of collection sites, standardized approaches to assess abundance and the careful definition of species, are all aspects of the observational studies that generate data for the analysis of species abundance (McGill et al., 2007).

A theoretical issue of much interest is the relationship between species abundance or population density and both average body mass and average body metabolic rate. Here there is disagreement among researchers. Theoretically it is usually postulated that a power law scaling exists between species density and body mass, and also body metabolic rate. Larger animals require more space per individual thus a lower population density is expected as body mass increases. This relationship, which implies a linear relationship on a log-log scale may hold generally or in the presence of specific sampling approaches or spatial assessments.

A factor affecting the interpretation of these types of measurements is the sampling approach used (Blackburn and Gaston, 1999). Note that even though technically the Bayesian perspective does not average over all possible datasets and often discussions do not formally address randomization and quality designs, it is always good practice to carefully collect data in a well planned and scientifically relevant setting. Frequentist and Bayesian perspectives should both reflect quality science and the sampling design and planned collection of data is very much a part of the applied science of a subject.

In the context of species abundance studies, the bringing together of data collected across a wide variety of catchment areas and species can make it difficult to argue that the overall dataset is an integrated whole that can be analyzed assuming a standard scaling of population density measures. There are often differences in the spatial aspect of these studies that impose differences in regard to the intensity of sampling for each study. Often the selection of how intensely to sample is a function of the size of the animal in question. Smaller animals require more intense sampling procedures (Blackburn and Gaston, 1999). Field studies also often reflect the practical properties of the different catchment areas to be used.

Other related issues affect the comparability of different catchment areas. These include the need to consider time and age in relation to the chosen catchment area. It may be the case that over time there is a richer set of species in a given catchment area reflecting the type of food supply and habitat available. While in other areas this may not be the case due to type of terrain or biome related factors. These would affect the sampling intensity in the different areas and thus perhaps imply heterogeneity of variation in average body size or average body metabolism across the set of catchment areas due to differences in sampling (Schemske and Mittelbach, 2010).

The stability of the climate and related stability of the biome also affect these considerations. Here human intervention often plays a role as changes in usage patterns by humans can alter greatly the habitats available to other species. On a larger scale as climate change and global warming occurs, the rise of sea levels will change or terminate many low lying habitats near the ocean.

Typically for the relationship between population density (D) and average body mass (M) we assume the following:

$$D = aM^{-b}$$

or on a logarithmic scale

$$\ln D = \ln a - b \ln M$$

The linearity of this relationship is questioned on the basis of data analysis and issues affecting sampling design. Interestingly while such power laws are theoretically supported, and graphically present, they are often not technically supported by formal hypothesis testing procedures. It is also worth noting that there are some statistical distributions that have a formal power law form (eg. Pareto distribution) and some that possess a power law relationship between their theoretical mean and variance (eg. exponential dispersion family of distributions). This last one can be seen as leading to a power law in the limit, a central limit theorem related justification for the onset of power laws, in a sense independent of the particular ecological population under study (Kendal and Jorgensen, 2011).

The relationship between population density (D) and average body metabolic rate (R) is given by:

$$D = aR^{b}$$

or

$$\ln D = \ln a + b \ln R$$

where theoretical arguments (Brown et al., 2002) can be made to set $b = 3/4$.
The basic statistical model here is initially:

$$\ln D = \ln a + b \ln R + \varepsilon$$

and depending on the question this model can be augmented. The ε_i are assumed to be *i.i.d.* $N(0, \sigma^2)$. The likelihood function is then available following standard regression modeling. In light of potential secondary and sampling related factors, this relationship may be assessed in a linear model controlling for additional variables.

Bayesian considerations here can reflect theoretical expectations and beliefs regarding the parameter b. This can be taken, in the case of average metabolic rate, to be for example a $N(3/4, \rho^2)$ distribution where the variance ρ^2 is given a non-informative distribution. If the use of a symmetric distribution for b about the theoretically justified mean is questionable then a skewed distribution can be used. The parameter a is essentially an intercept on a log scale.

The observed likelihood function will re-weight and adjust the prior mean values to reflect the observed data, subject to the assumption of the power law relationship and the chosen prior distributions for the parameters. We can then compare various b values of interest by examining posterior odds ratios. Robustness can be assessed by altering priors and repeating the analysis. Note that where there is a strong disbelief in the theoretical value for b, the assumption of, for example, $b \sim N(0.25, \rho^2)$, might be a more appropriate prior density.

In models where the basic relationship is augmented, for example using polynomial terms to deal with a potential nonlinear pattern, or controlling for environmental and/or sampling related variables, polynomial regression or linear models with covariates can be applied. Note that if we use a polynomial fit in the $\log - \log$ model, the theoretical value for b and associated prior densities may require modification as we will have additional parameters to model. In real-world problems, as we consider various models, the relevant theory may also require different interpretation and may lead to a different set of prior densities. This depends on the context.

The assumption of $b = -3/4$ for the species abundance-average body mass relationship has a similar detailed theoretical background and discussion. However studies exist that have found b values of -1.40, -.88, -.93 depending on the species in question and sampling approach taken. The relationships between species abundance and average body mass and metabolism as a fundamental aspect of ecological and biological relationships remain the subject of much study.

8.3   Data

The data here are simulated based on a real-world study examining bird species (Silva and Downing, 1995; Silva et al., 2001) abundance data drawn from a variety of studies collected over an extended time frame and integrated

TABLE 8.1
Variable Listing

Name	Description
Species Abundance	0 - 100, Continuous
Species Abundance High-Low	Discrete
Average Body mass	Continuous
Average Body Metabolic rate	Continuous
Recent changes in nearby human activity	Categorical
Year	Discrete
Sampling Intensity	Low, Medium, High

into a large meta-analytic database. Table 8.1 lists the variables available for the analysis here.

8.4   Specific Aims, Hypotheses, Models

In the analysis developed here we primarily examine whether power law relationships exist between population density and average body size and population density and average body metabolic rate. From the simulated database we study the hypothesized body size and body metabolism - species abundance relationship, along with potential evidence of nonlinearity subject to caveats of underlying issues regarding scaling and sampling. Initial basic analysis includes simple regression analysis as well as polynomial linear regression. Summary tables present mean, median, sample size and quartiles. Graphics are also provided as appropriate. Dendrogram analysis is also carried out, as well as simple PCA based data analytic clustering methods where appropriate.

Note that clustered sampling designs and the need for random effects interpretations can also be relevant. These reflect the use of data collection based on catchment area and collection center and the area covered by the collection center. Often the number of collection centers or catchment areas is limited and the chosen catchment areas are seen as an observed set of areas out of a broader population of potential catchment areas. This can justify use of the random effects interpretation on the observations selected in such designs. As well if individual animals are marked and tracked there is a need for random effects in the form of repeated measures analysis (this is not examined here).

Note that the relationships of interest can also be given a different basic model, for example using a threshold model format to analyze a continuous response we have re-scaled (above or below the median value for example). A logistic regression can be applied if such a threshold level can be defined to relate population density above a specific level to average body size and average body metabolism.

TABLE 8.2

Data Summary

Variable	n	Mean	Std Error	Q1	Median	Q3
Species Abundance	100	0.811	0.121	-0.05	1.14	1.875
Average Body Mass	100	0.409	0.146	-0.958	0.24	1.425
Average Body Metabolism	100	0.503	0.156	-0.89	0.67	1.768

For each model Bayesian priors are developed and interpreted, based on non-informative structure or drawn from existing knowledge. Maximum likelihood estimators along with standard p-values and confidence intervals are reported and Bayesian posterior credible regions. Predictive distributions are studied as appropriate. An integrated summary is then developed, examining the overall data analytic and likelihood based picture provided by the likelihood, *m.l.e.*, information matrix, posterior and predictive densities.

Specific aims and hypotheses for examination include:

1. Validate the fit of the basic power law relationships overall. Note the estimation of b is also important here as it has theoretical interpretation.

2. Does a nonlinear relationship hold between population density and body size and average metabolic rate? Here loess and polynomial models are applied to address this question.

3. Is the fit of the power law relationships validated across subgroup and sampling intensity levels? Is the b parameter similar across subgroups?

8.5   Analysis and Interpretation

8.5.1   Data Analysis

The dataset is a simulation based on the data and analysis applied in (Silva and Downing, 1995; Silva et al., 2001). The data are available as supplementary materials. Some basic data summaries for individual variables are given in Table 8.2 (on a log scale).

Scatterplots are given in with fitted regression lines. These show a basic power law relationship between species abundance and both average body mass and average body metabolic rate.

A dendogram is also used to further examine potential clustering patterns in the data as well, especially in regard to basic design issues and the comparability of the studies included in the dataset. Here we use the species abundance measure and the other two continous variables to generate possible clustering

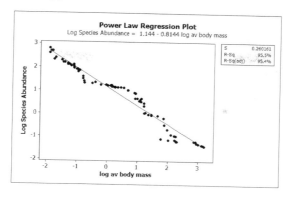

FIGURE 8.1
Body Mass Power Law Regression Plot

in the dataset. No clear pattern is observed. We might also repeat this analysis for various subgroups of interest to examine the analysis for stability across subgroups.

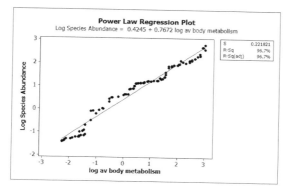

FIGURE 8.2
Body Metabolism Power Law Regression Plot

Homogeneity is an issue and may not be an acceptable assumption as sample variations across subgroups may vary. For example the variation across the three sampling intensity groups. Here the residual plots for these subgroups are similar to the overall group patterns. The regression plot for species abundance to average body mass is shown for sampling group 3.

Note as a caveat to inferences that the data are not a random sample drawn in a manner to limit bias or non-comparable sampling intensities. This is typical of large meta-analytic databases.

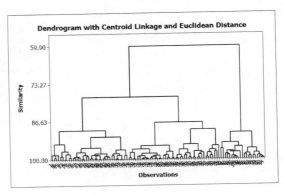

FIGURE 8.3
Dendrogram Plot

8.5.2   Likelihood Function

Assuming a normal error distribution here we have as our initial overall likelihood function:

$$L(\boldsymbol{\beta}, \sigma^2 | \mathbf{y}) = c \cdot (1/2\pi\sigma^2)^{n/2} \exp(-(1/2\sigma^2) \sum_i (y_i - \mu)^2)$$

where $\mu = \beta_0 + \beta_1 x_{1i} + \cdots + \beta_p x_{ip}$ in a regression setting.

In the frequentist setting the least squares and ANOVA based analysis is closely reflected in the likelihood based analysis with the least squares estimators of the β_j also being the *m.l.e.* estimators. Slight differences in estimates of σ^2 are minor in large samples.

If more detailed likelihood analysis is pursued in regard to the power law aspect, a more appropriate model may be the exponential dispersion family, an extension of the exponential family of densities discussed earlier in Parts II and III. This is a source of probability densities where the mean and variance are in a power law relationship, giving the likelihood:

$$\prod_{i=1}^n f(y_i; \theta, \lambda) = \prod_{i=1}^n a(\theta, \lambda) \exp[\lambda\{y_i\theta - \kappa(\theta)\}]$$

where $\lambda > 0$, θ lies in an interval on the real line, $a(\theta)$ and $\kappa(\theta)$ are functions of θ. The mean of y_i is given by $\kappa'(\theta)$ and the dispersion parameter is given by $\sigma^2 = 1/\lambda$. See for example (Kendal and Jørgensen, 2011).

If the response can be rescaled onto a High/Low scale this implies modeling the probability of "High" as a function of the chosen explanatory variables using the logistic function. The fitted probability or dose response function is then given by the general form $\hat{p} = \exp(x)/[1 + \exp(x)]$. A generalized linear model can then be applied.

The likelihood function relevant to logistic regression is given by:

$$L(\boldsymbol{\theta} \mid \mathbf{y}) = c \cdot \prod_{i=1}^{n} \left[\frac{\exp(\sum_{j=0}^{k} \theta_j x_{ij})}{1 + \exp(\sum_{j=0}^{k} \theta_j x_{ij})} \right]^{y_i} \left[1 - \frac{\exp(\sum_{j=0}^{k} \theta_j x_{ij})}{1 + \exp(\sum_{j=0}^{k} \theta_j x_{ij})} \right]^{1-y_i}.$$

Frequentist based analysis typically determines the values of θ_j giving the maximum value for the likelihood or log-likelihood function. These are then assessed in terms of probability using large sample arguments.

8.5.3 Likelihood Frequentist Analysis

As the primary goal here is to model the power law responses, we see that the graphics do not contradict the assumption of a linear model here. Non-parametric or local likelihood based loess models can be run here to detect additional structure but in this particular dataset there does not seem to be much structure in these simple regression lots other than a possible nonlinear wave.

Polynomial regression is fit (quadratic and cubic, cubic shown) and some of these are significant, however compared to the initial linear fit, the R^2 values are similar, at the approximately 95% level throughout. In this type of setting, the principle of Occam's razor is useful; take the simplest explanation. This would be the linear regression supporting the assumption of a power law.

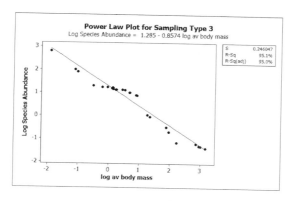

FIGURE 8.4
Body Mass Power Law Regression Plot for Sampling Type 3

This allows for the consideration of the slope of the simple linear regression lines. We obtain for the species abundance to average body mass $-0.814 \pm 2(0.01795)$, and for the species abundance to average body metabolism $0.7672 \pm 2(0.01433)$ as approximate 95% confidence intervals. As noted earlier these have theoretical implications.

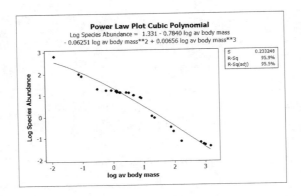

FIGURE 8.5
Body Mass Power Law Regression Plot with Cubic Polynomial

TABLE 8.3
Regression Summary

Regression Equation	Log Spec Abund = 1.02 - 0.821 log av body mass + 0.065 sampling			
	Coefficient	Std Error	T Statistic	p-value
Constant	1.02	.0695	14.63	0.001
Log Average Body Mass	-0.821	.0179	-45.65	0.0001
Sampling Intensity	0.065	.033	1.97	0.052
R^2, MSE	.956, 0.066			

A sampling intensity measure here is categorized into low, medium, high and this is used to group the data. The power law relationship to average body mass was examined across these groups. The fitted b value is found to be fairly stable at 0.80. A plot is given of sampling intensity group 3 in the section above.

Using a broader linear model to examine potential effects of sampling intensity as a control variable gives no significant change. The least squares analysis provides the following results for log body mass (log body metabolism has a similar sampling intensity effect (p-value = 0.052, $R^2 = 0.968, MSE = 0.048$). See Table 8.3.

8.5.4   Likelihood Bayesian Analysis

The goal in any analysis is the collection of useful and interpretable information and understanding. The Bayesian approach provides a modified perspective on the likelihood function. This requires a prior density that models existing belief regarding the parameter θ. This then allows us to update our

TABLE 8.4
Ninety-Five Percent Confidence and
Credible Intervals

Prior Assumption	b
Non-Informative	$(-.848, -.780)$
$b \sim N(-0.60, 0.0001)$	$(-.824, -.761)$
$b \sim N(-0.80, 0.0001)$	$(-.861, -.781)$
m.l.e. **No prior**	$(-.850, -.778)$

beliefs using the likelihood function and Bayes theorem, which is a direct corollary of conditional probability.

For the species abundance - average body mass relationship, we assume a Normal prior for b and a non-informative density for $\ln a$, along with the standard inverse gamma for σ^2. The marginal posterior density is obtained and the associated 95% HPD credible regions for b given in Table 8.4 along with *m.l.e.* based results.

If we do not believe the theory that $b \sim N(-.75, 0.0001)$ we can use the prior setting to register our disbelief in a practical way that weights the model and results directly. For example we might use the priors $b \sim N(0.60, 0.0001)$ or $b \sim N(0.8, 0.0001)$.

8.5.5 Analysis/Integration/Comparisons

As we move from data analysis to assumed likelihood and likelihood based inference to the Bayesian perspective on the likelihood function and other Bayesian inferential concepts, we are slowly making more mathematical and structural assumptions regarding the scientific problem.

The frequentist analysis is based on the properties of the likelihood. In the case of normality and larger samples this extends to least squares. Here the frequentist analysis gives support to the chosen parameteric linear models and also does not contradict the use of the overall estimates and tests as clusters observed do not seem very distinct.

We obtain for the species abundance to average body mass $-0.814 \pm 2(0.01795)$, and for the species abundance to average body metabolism $0.7672 \pm 2(0.01433)$ as approximate 95% confidence intervals. Both are drawn from regression settings that show a basic power law shape. The theoretically supported slope of -.75 in regard to body mass is not supported here, while the .75 slope value for body metabolism is supported.

The Bayesian prior here reflects available past studies. The posterior reflects our updated beliefs when we have processed likelihood and prior through Bayes theorem and obtained the posterior density. It is flexible in the sense of reflecting the personal beliefs and chosen prior density of the researcher. Here as there is a developed scientific literature and results the chosen priors reflect past theory and findings.

In terms of basic results, the Bayesian credible intervals for b here overlap with the frequentist interval and the similarity of results supports the basic power law model, even in the presence of priors not supportive of the $b = -.75$ theoretical value for average body mass. Similar results hold here for average body metabolism. Thus practically here the chosen priors do not alter the basic conclusions. What do we learn from the Bayes setting here is the stability of the results with regard to the formally modeled levels of belief and disbelief in the existing theory. The predictive density is of interest as it provides truly probabilistic intervals for future values where the stochastic process is well modeled by the likelihood.

In the Suggested Exercises below, more detailed assessments are suggested, reflecting the Bayes learning model perspective.

8.6   Suggested Exercises

In real-world settings the number of hypotheses to be considered can be wide ranging. They reflect the design and observed data, but modern statistical models can be adjusted in many ways to account for issues such as missing data, clustering, unstable baseline measures.

1. Consider the use of Bayes factors in assessing the hypotheses considered above. Look at the cutoffs given in Chapter 3 to evaluate. Use Bayes factors to compare the following models for species abundance: 1 (average body mass) and 2 (average body mass, average body metabolic rate, sampling intensity). Compare this to the frequentist approach using partial F-tests for the same models. Does the use of a different prior affect the similarity of the approaches here?

2. Take the observed population density values and recode as 1 (if greater than 50%) and 0 (if less than 50). Use a logistic regression and simple dose response type model to relate the probability of a "high" population density to the average body mass. Comment on the shape of the logistic curve. Do the same for average body metabolism. Note that by reducing the respnse to 0 or 1 we are no longer using the actual response and are losing accuracy and thus information in return for accessing the dose-response format.

3. Take a median split on the species abundance, average body mass and average body metabolism values. Derive three two-by-two contingency tables. Find and interpret the odds ratio. Perform a chi-square or Fisher exact test to assess whether the variables are associated. Use a binomial model to express the likelihood here as well as priors and possible posteriors.

4. Discuss the use of the exponential dispersion family which assumes a power relationship between mean and variance. What types of priors would be useful with this family of likelihoods? Given that central limit theorems here will lead to convergence to a power law relationship between mean and variance, does this imply that the power law relationship in the species abundance setting is a result of central limit theorems having little to do with the science involved?

5. The Bayesian perspective is based primarily on probabilistic summaries; the posterior densities of interest. Differences in the data overall and within subgroups can be assessed with regard to inferences for a given parameter θ_j by measuring the change in the marginal posterior $p(\theta_j|data)$ for a given prior density. One measure that can be used for this is a version of the entropy measure; $E_{i,k} = \int p_i(\theta_j|data) \log(p_i(\theta_j|data)/p_k(\theta_j|data))d\theta_j$ where k represents the overall data and i the subgroup of interest. If the $E_{i,k}$ values are fairly constant the analysis regarding θ_j can be seen as stable in terms of the marginal posteriors for θ_j.

6. Define the Simpson's paradox. Discuss where the Simpson's paradox might occur in the species abundance dataset examined here. Is there evidence of the paradox in regard to the species abundance - body size relationships found here? Are the results found regarding linearity and the power law relationship ? Consider both frequentist and Bayesian perspectives and measures here.

7. The intensity of sampling may vary across catchment area and study site. It has been argued that this affects interpretation of the power law relationship. If this effect expresses in terms of heterogeneity (non-constant variance) discuss how to adjust the models here to allow for non-constant variation (for example the use of a weighted least squares model). Express this in terms of likelihood and discuss how to choose priors in this setting.

8. How extreme would the chosen prior for b have to be to give a less than 50% overlap of the likelihood frequentist and likelihood Bayesian results?

8.7 Bibliography

[1] Blackburn T.M. and Gaston K.J. (1999). The Relationship Between Abundance and Body Size: A Review of the Mechanisms. *Advances in Ecological Research* 28, p. 181–210.

[2] Brown J.H., Gupta V.K., Li B.-L., Milne B.T., Restrepo C. and West G.B. (2002). The Fractal Nature of Nature: Power Laws, Ecological Complexity and Biodiversity. *Phil. Trans. R. Soc. Lond. B* 357, p. 619–626. DOI 10.1098/rstb.2001.0993.

[3] Kendal W. S. and Jørgensen B. (2011). Tweedie Convergence: A Mathematical Basis for Taylor's Power Law, 1/f Noise, and Multifractality. *Physical Review E (Statistical, Nonlinear, and Soft Matter Physics)* 84(6), 066120-1-066120-10doi: 10.1103/PhysRevE.84.066120.

[4] McGill B.J. et al. (2007). Species Abundance Distributions: Moving Beyond Single Prediction Theories to Integration Within an Ecological Framework. *Ecol. Lett.* 10, p. 996–1015. (doi:10.1111/j.1461-0248.2007.01094.x).

[5] Schemske D.W. and Mittelbach G.G. (2017). Latitudinal Gradients in Species Diversity: Reflections on Pianka's 1966 Article and a Look Forward. *The American Naturalist.*

[6] Silva M. and Downing J.A. (1995). The Allometric Scaling of Density and Body Mass: A Nonlinear Relationship for Terrestrial Mammals. *American Naturalist* 145, p. 704–727.

[7] Silva M., Brimacombe M., Downing J.A. (2001). Effects of Body Mass, Climate, Geography, and Census Area on Population Density of Terrestrial Mammals. *Global Ecology & Biogeography* 10, p. 469–485.

9

Soil Erosion in Relation to Season and Land Usage Patterns

9.1 Science

Soil science, the study of soil maintenance and erosion, agricultural use of land, farming and fertilization practices has a long history. In relation to statistical methods, the analysis of crop rotation was one of the first areas of application of ANOVA (Fisher, 1921). Since then the analysis of soil use and erosion patterns has become much more advanced, forming a key aspect of ecological research. Climate change and related phenomena have a direct impact on soil health and agricultural productivity.

Many agricultural plants may have specific suggested planting rotations, designed to limit erosion and maintain soil health, approaches that are growing in interest and study as the large scale application of pesticides and chemical control of insect and blight is becoming less acceptable in the face of demand for organically grown foods.

In terms of a formal definition, "Soil erosion is the wearing away of the land surface by physical forces such as rainfall, flowing water, wind, ice, temperature change, gravity or other natural or anthropogenic agents that abrade, detach and remove soil or geological material from one point on the earth's surface to be deposited elsewhere. Soil erosion is a natural process that can be exacerbated by human activities," (Bosco et al., 2009).

The modeling of environmental processes require the judicious application of statistical sampling ideas and practical, robust models. Soil erosion is an example of this and often reflects a large number of variables as well as high levels of heterogeneity when modeling this data. Soil erosion may reflect natural ongoing processes, for example those affecting beachfront areas in coastal settings. It may also reflect specific agricultural practices, for example the planting of specific types of plants requiring harvesting methods that erode soil integrity. Obtaining a stable baseline set of measurements against which to compare and assess ongoing erosion is a challenge and when available is often expressed in the form of an index value, taking a specific year as the initial year of measurement to compare against.

For example, in northern potato growing climates, the effects of snow and snow melt runoff on soil retention is profound in years when potatoes are grown. These interact with harvesting practices that often deplete the soil. This is also affected by the physical characteristics of the land in question, the inclination or tilt of the land, the weather conditions and the type of crop rotation employed. The seasons during which harvesting takes place also impacts soil erosion. Typically a three year crop rotation is employed to preserve the integrity and fertility of the topsoil.

As noted in the literature (Mullan, 2013), climate change seems to be affecting the moisture-holding capacity of the atmosphere, modifying the frequency, duration and intensity of rainfall. Increases in temperature and related CO_2 concentrations can lead to changes in plant biomass, increasing erosion rates. Farmers may adapt the timing of agricultural operations, changing planting and harvesting dates with new crops, factors to consider in regard to measuring patterns of soil erosion. All of these considerations affect the use of statistical methods.

The study of soil erosion is usually conducted in a set of carefully chosen plots for which fairly complete information is available and verifiable. Often in countries with large agricultural sectors, government and corporate funded research farms are maintained for the study of soil and crop maintenance. Large scale agricultural surveys conducted every five or ten years may also provide background information. Spatial graphical analysis using technology such as drones or satellite imagery may also provide useful overviews, especially for spatial analysis of changes over time or the results of extreme weather conditions such as drought or flooding.

Study of soil management also extends to questions addressing the interconnectivity of many components of the agricultural and environmental setting. For example where pesticides are used, nitrogen runoff from agricultural fields into nearby rivers and lakes may result in large scale die-offs of fish and other aquatic life. Accurately measuring the threshold levels of such incidents is a difficult and resource consuming effort. These have obvious political and economic ramifications. Some efforts to develop such assessments can be found in Edwards et al., (2000).

9.2 Data

The data here are based on and simulated from a soil science study conducted in PEI, Canada (Edwards et al., 1998, 2000). The variables are measured at specific sampling sites with soil erosion over a three year period the outcome of interest. Average rainfall, average snowfall, crop rotation, inclination of the site are related to outcome. The response of interest is also rescaled into an erosion index where the soil baseline level is taken at the beginning of a specific planting rotation and associated year.

In northern settings snowfall and land incline are factors in soil erosion. In

TABLE 9.1

Variable Listing

Name	Description
Soil erosion index level (y)	Continuous
Crop rotation	Yes/No
Incline/tilt	Continuous
Monthly rain level	Continuous
Monthly snow level	Continuous
Monthly average temperature	Continuous

an environment close to the ocean where there are multiple rain episodes in the January-April period, this is the time of maximum erosion as the melting ice has substantial weight and can greatly affect soil erosion when related to the incline or tilt of the field in question. Seasonal run-off can also be an issue, but the nitrogen runoff that can lead to fish die-offs in nearby rivers is limited by the absorption of the nitrogen fertilizer into lower levels of the soil (Edwards et al., 2000).

For each measurement site a standardized method of measurement for soil erosion and other variables was employed. Given the variety of weather pattens and their constant changes, averages are reported and non-constant variation may be an issue. A number of non-potato rotation fields are included to provide a "control" group in regard to naturally occuring soil erosion. Table 9.1 gives the variables collected for each site included in the analysis here.

9.3 Specific Aims, Hypotheses, Models

The basic aim here is to provide an overview and model of soil erosion occurring over a 6-year period in fields currently in potato production. Note that soil erosion itself is measured here as a continuous measure, with changes defined for April-January measurements in each year. Some fields follow specific three-year rotation (alfalfa, wheat, potatoes), some do not. In regard to specific hypotheses, the relationship between soil erosion and several variables; incline/tilt, rotation, crop type, average rain level and average snow level, are examined individually and within larger regression models.

Questions that arise include:

1. Are there any clustering or correlation patterns among variables and individual observations? Do these alter in relation to higher levels of soil erosion?

2. Are specific factors such as incline, rain and snow melt related to soil erosion? Estimate this relationship.

3. Do these relationships hold up in the presence of control variables and in subgroups?

Data analysis includes data tables presenting mean, median, standard deviation, median, range and IQR values. Graphics include boxplots, dendrograms, histograms and various scatterplots. PCA is applied where appropriate. These graphics are carried out both overall and within chosen subgroups, giving the opportunity for comparison. Given the variety of fields and management practices, Simpson's paradox is a worry here and the various analyses are sometimes repeated for smaller subgroups of the overall set of sampled fields. As a simple measure of heterogeneity, sample variance across subgroups are compared.

Models that can be employed include linear and logistic regression models, random effect models, linear models allowing for heterogeneity, as well as the modeling of patterns through time. For categorized versions of the variables using median based thesholds and related contingency table analysis is conducted with measures such as relative risk and odds ratios estimated and interpreted.

Maximum likelihood and least squares based estimators and confidence intervals are obtained. For the Bayesian setting marginal posteriors for regression parameters, odds ratios and relative risk measures are found. The Bayesian perspective is often not employed in such settings, as randomization of various crop patterns and soil types typically require a design of experiments focus rather than just likelihood and the analysis of observed data. That said, there are some theoretical expectations and previous studies from which to draw insight in the selection of prior baseline beliefs and expectations and Bayesian approaches often provide an approach to averaging complicated models. Note that due to variation, priors may not be relevant in years with extreme change and these have recently become more common. A more robust approach to initial prior selection can be pursued and conjugate priors applied where justified.

9.4   Analysis and Interpretation

9.4.1   Data Analysis

The dataset is a simulation based on a study conducted in PEI Canada (Edwards et al., 1998, 2000). The data are available as supplementary materials. Basic data summaries for individual variables are given in Table 9.2.

As the goal here is to model the soil erosion response with a linear model, we use graphics to support the idea that the data do not contradict the assumption of a linear model here. Two matrix plots are given, one using loess fits and the other linear regression.

TABLE 9.2

Data Summary

Variable	n	Mean	Std Error	Q1	Median	Q3
Soil Erosion Index	60	6.403	0.169	5.6	6.8	7.375
Average Rain	60	13.943	0.289	11.8	14.1	15.0
Average Snow	60	61.09	1.31	53.08	65.0	67.78
Tilt	60	5.10	0.191	3.95	5.0	6.35
Temperature	60	5.183	0.183	4.0	5.0	6.0

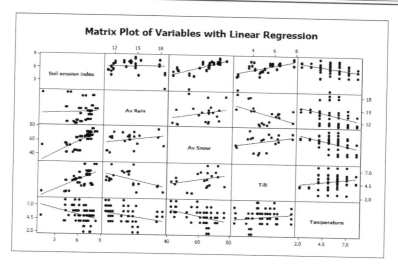

FIGURE 9.1

Multiple Scatterplots

A dendrogram is used to further examine potential clustering patterns in the data providing visualization of the overall correlation pattern. Note these types of summaries do require basic linear relationships between the variables.

Here we see that the data do not contradict basic assumptions regarding the applicability of the linear model here. The correlation matrix of the continuous variables reflects the pattern above. See Tables 9.3 and 9.4.

Homogeneity across rotation group in regard to soil erosion is supported with standard deviations $(1.25, 1.21)$ for rotation group 0 and 1 respectively.

9.4.2 Likelihood Function

Assuming a normal error distribution here as well as a linear regression structure we have as our initial overall likelihood function:

$$L(\boldsymbol{\beta}, \sigma^2 | \mathbf{y}) = c \cdot (1/2\pi\sigma^2)^{n/2} \exp(-(1/2\sigma^2) \sum_i (y_i - \beta_0 - \beta_1 x_{1i} - \cdots - \beta_p x_{ip})^2)$$

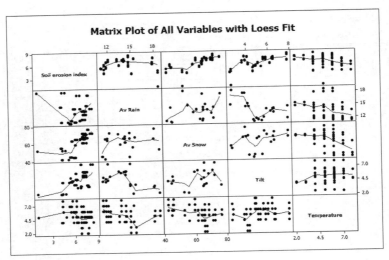

FIGURE 9.2
Multiple Scatterplots with Loess Fit

TABLE 9.3
Correlation Matrix

	Soil Erosion	Av Rain	Av Snow	Tilt
Av Rain	.046	—	—	—
Av Snow	.728	.272	—	—
Tilt	.551	-.498	.325	—
Temp	-.321	-.352	-.427	.188

If a student-t distribution is thought to be more useful, typically for potential outliers (say 5 degrees of freedom) the likelihood can be written:

$$\prod_{i=1}^{n} f(y_i; \alpha, \lambda, \mu) = \prod_{i=1}^{n} c[1 + \frac{\lambda}{\alpha}(y_i - \mu)^2]^{-(\alpha+1)/2}$$

$$c = \frac{\Gamma((\alpha + 1/2))}{\Gamma(\alpha/2)\Gamma(1/2)} \left(\frac{\lambda}{\alpha}\right)^{1/2}$$

with $\lambda > 0$ and $\alpha > 0$. The α parameter is termed the degrees of freedom (so $\alpha = 5$ for the t_5 distribution). In the setting of multiple regression set $\mu = \beta_0 + \beta_1 x_{1i} + \cdots + \beta_p x_{ip}$. Note that the question of why such a distribution is not typically used in frequentist settings relates primarily to the use of ANOVA, where the orthogonal decomposition uses zero correlation and normality to achieve an independent set of squared lengths used to form F−statistics for example. This is not the case with the t distribution in smaller samples.

TABLE 9.4

Regression Summary

Regression Equation	Soil = 0.60 + 0.055 Av Rain + 0.061 Av Snow + 0.42 Tilt - 0.16 Temp			
	Coefficient	Std Error	T Statistic	p-value
Constant	0.596	1.237	0.48	0.632
Average Rain	0.055	0.062	0.88	0.383
Average Snow	0.061	0.014	4.35	0.001
Tilt	0.42	0.099	4.26	0.001
Temperature	-0.16	0.086	-1.90	0.063
R^2, MSE	.668, 0.614			

9.4.3 Likelihood Frequentist Analysis

The analysis is primarily based on a linear model and ANOVA based decomposition of variation. The model parameters are estimated throughout using *m.l.e.* or least squares based estimators and standard errors.

The overall regression equation, leaving out crop rotation which was not significant, is given in Table 9.4.

Based on the assumed likelihood and observed data we have as our estimates and 95% confidence regions for tilt $(0.42 \pm 2(0.0995))$, average snowfall $(0.061 \pm 2(0.014))$, average rainfall $(0.055 \pm 2(0.062))$, temperature $(-0.16 \pm 2(0.086))$. Further there is a significant relationship between the soil erosion index and both average snowfall and tilt of sampling site when all variables are considered in a linear model (leaving out rotation).

Residual plots for the reported overall linear model support the choice of likelihood in the sense that the pattern in the residual plot does not contradict the assumed likelihood.

From a strict likelihood principle perspective, examining residuals is questionable, but it is essential, along with initial scatterplots, to ensure that the assumption of linearity holds. If it does not continuing to use a linear model would incur mis-specification errors.

9.4.4 Likelihood Bayesian Analysis

We now place the likelihood in the Bayesian context, using it to update our beliefs as expressed in a chosen prior density. Assuming normal diffuse priors for linear parameters and an inverse gamma distribution for σ^2, we can generate the required marginal posterior densities for all parameters of interest. From past experience absolute parameter values that differ from zero are expected for all variables. For this simulation we set the prior means at one standard error below the observed coefficient values. The associated 95% HPD credible regions are given in Table 9.5.

Note the regions do not alter greatly under the various prior assumptions

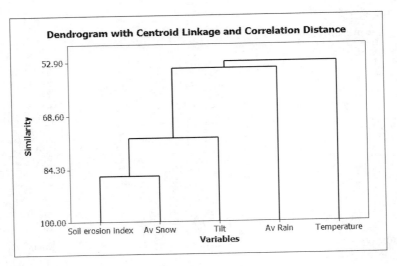

FIGURE 9.3
Dendrogram Plot

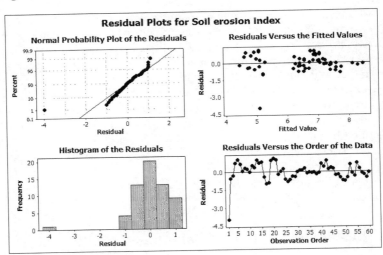

FIGURE 9.4
Residual Plots

TABLE 9.5
Ninety Five Percent Confidence and Credible Intervals

Prior Assumption	Av Rain	Av Snow	Tilt	Temperature
Non-Informative	$(-.061, .172)$	$(.029, .084)$	$(.210, .702)$	$(-.181, .005)$
Normal	$(-.057, .163)$	$(.024, .079)$	$(.204, 0.692)$	$(-.179, .001)$
m.l.e. No prior	$(-.069, .179)$	$(.033, .089)$	$(.222, 0.718)$	$(-.188, .012)$

and the basic inferential results hold where $\beta_j = 0$ is the reference value. Further extensions of Bayesian analyses are outlined in the suggested exercises.

9.4.5   Analysis/Integration/Comparisons

The analysis here proceeds from data analysis to assumed likelihood and likelihood based inference to the Bayesian perspective on the likelihood function, slowly making more mathematical and structural assumptions regarding the scientific problem itself as they are supported by the data and basic models.

Here the data analysis gives support to the chosen parameteric models and also does not contradict the use of the overall estimates and tests as the clusters observed do not seem very distinct. The correlation and clustering observed are reflected in the linear model fit to the data.

The frequentist analysis is based on the properties of the likelihood. Here the soil erosion index is seen to be related to average snowfall and tilt of the sampling site. The model is not particularly useful ($R^2 = 0.67$) and reflects a fairly large level of variation and several outlying values.

The use of Bayesian methods in the linear model setting has a very long history. See for example Lindley and Smith (1972). As MCMC based methods have been improved the variety of prior densities available have greatly increased as have the types of likelihood functions that can be considered.

Considering that the credible and confidence regions for average rainfall and tilt both do not contain zero, we have a model-data supported association with the soil erosion index from Bayesian and frequentist perspectives. This may alter if stronger prior assumptions are employed and justified, if additional assumptions are added to the model, for example non-constant variation or a likelihood that reflects the presence of outlying values, for example the Student-t distribution.

Note that while past studies are generally useful in determining what ranges and other basic properties to expect, often the variations across crop type and soil collection site is non-trivial in soil science studies and typically would lead to a large hierarchical model with each component of the model supported by relatively small sample sizes. It is always helpful to choose a proper prior in such settings.

9.5   Suggested Exercises

1. The models considered here have focused on assessing soil erosion in regard to specific conditions. Assuming a normal likelihood and independent normal priors obtain and examine the Bayesian predictive distribution in regard to soil erosion.

2. Use independent normal priors and Bayes factors to compare two

models of soil erosion using (i) crop rotation and inclination (ii) rotation, crop type, inclination. Compare the results to those obtained using standard partial-F tests (log-likelihood based) from the frequentist perspective. Repeat the Bayes factor calculations using a non-informative prior. Does the assumed prior affect the similarities observed?

3. Use a median based threshold for soil erosion and create a threshold variable (1/0) if above or below the median. Examine the data in these two groups using dendrogram plots and contingency tables. If relatively similar, model the probability that soil erosion is high using a logistic regression model with random effect from both frequentist and Bayes perspectives. Compare and discuss.

4. The Bayesian perspective is based primarily on probabilistic summaries; the posterior densities of interest. Differences in the data overall and within subgroups can be assessed with regard to inferences for a given parameter θ_j by measuring the change in the marginal posterior $p(\theta_j|data)$ for a given prior density. One measure that can be used for this is a version of the entropy measure; $E_{i,k} = \int p_i(\theta_j|data) \log(p_i(\theta_j|data)/p_k(\theta_j|data))d\theta_j$ where k represents the overall data and i the subgroup of interest. If the $E_{i,k}$ values are fairly constant the analysis regarding θ_j can be seen as stable in terms of the marginal posteriors for θ_j. Do this for a model examining soil erosion in relation to inclination and rotation. Examine the stability in marginal posteriors using $E_{i,k}$ across crop type.

5. Use the Student-t(5) distribution to obtain the likelihood function in the Bayesian analysis. Discuss any differences in the observed marginal posteriors that arise from using a distribution that allows for a larger variation to address potential outliers.

9.6 Bibliography

[1] Bosco C., Rusco E., Montanarella L. and Panagos P. (2009). Soil Erosion in the Alpine Area: Risk Assessment and Climate Change. *Studi Trent. Sci. Nat.* 85, p. 117–123.

[2] Edwards L., Bernsdorf B., Pauly M., Burney J.R., Satish M.G., Brimacombe M. (1998). Spatial Interpolation of Snow Depth and Water Equivalent Measurements in Prince Edward Island. *Canadian Agricultural Engineering* 40, p. 161–168.

[3] Edwards L., Burney J., Brimacombe M., Macrae A. (2000). Nitrogen Runoff in a Potato-Dominated Watershed Area of Prince Edward Island, Canada, in *The Role of Erosion and Sediment Transport in Nutrient and Contaminant Transfer (Proceedings of a symposium held at Waterloo, Canada, July 2000).* IAHS Publ. No. 263.

[4] Fisher R.A. (1921). Studies in Crop Variation. *J. of Agric. Sci.* 11, p. 107–35.

[5] Lindley D.V. and Smith A.F.M. (1972). Bayes Estimates for the Linear Model. *J. Roy. Statist. Soc. B. 34*, Vol 1, p. 1–41.

[6] Mullan D. (2013). Managing Soil Erosion in Northern Ireland: A Review of Past and Present Approaches. *Agriculture* 3, p. 684–699; doi:10.3390/agriculture3040684.

10

Case Studies in Biology

The application of statistical models to biological study, once referred to as biometrics, has a long history. Models that have focused on toxicology, genetics and epidemics have been in use for almost 100 years and the statistical methods developed to analyze such data reflect most of what we might call standard statistical and biostatstical methods, including least squares, random effects, likelihood based methods, nonlinear regression, generalized linear models and ANOVA.

Biology itself is a vast area of scientific endeavor, focusing on many aspects of animal, human, viral and other forms of life. Studies in this field can be based on precisely calibrated laboratory data, or large scale databases that require data analytic methods to be investigated and themselves challenge standard statistical intuition. There is typically biological theory in these settings that may be helpful in guiding the development of likelihood functions and the selection of prior densities.

Recently the incorporation of genomics has greatly expanded the opportunity to use statistical and mathematical models to discover and summarize various patterns in observed data. In particular clustering related methods are of interest and the need to relate outcomes to a very large number of potential explanatory variables, often greater than the number of subjects in the study. Large databases may also be developed without the use of statistical design principles to guide the data collection. Data structures such as networks and hierarchical growth models can also be applied in these contexts.

The models employed can often have a non-linear aspect to them. While this has typically been minimized through the careful selection of underlying mathematical models, it can present a challenge. It is also important to note that whatever perspective is employed, Bayesian or frequentist, the properties of good design do not alter. Careful randomization, replication whenever possible to examine the goodness of fit of assumed models, the collection of representative samples, designs to minimize confounding and bias, all represent good science and good statistical practice.

Here we examine the modeling of antibiotic resistant TB in relation to specific genes, the use of temperature level and dosage in the controlling of the onset of sea lice in fish farming, and large database clustering and linear modeling in relation to mouse liver cancer genomics.

10.1   Immunity and Dose Response in Relation to Aquaculture

10.2   Science

Fish farming is an area of aquaculture that has seen intense development over the past several decades. Fish farms have become common and require large amounts of relatively clean water with carefully controlled feed and antibiotic use as well as restriction on movement, often to limit the interaction among genetically bred groups of fish and wild type. These include species such salmon and rainbow trout. Growth in the size of these farms is typically limited by the supply and maintenance of clean water. Pollutants such as algal blooms can multiply at an exponential rate, using up nutrients and causing large scale die-offs. Water purification is a key component of fish farms and often vats or cages are placed in rivers or in the sea.

The risk of infections by parasites such as fish lice, fungi, intestinal worms, bacteria and protozoa is similar to those arising in animal farms and sea lice often cause deadly infestations in farmgrown fish. Sea lice latch onto the skin and limit fish growth. Large numbers of densely populated, open-net farms can create concentrations of sea lice and many young fish may not survive.

In the context of fish farming, dosing individual fish is challenging as they may not all consume similar levels of feed. Thus treatments are often administered in both feed or in the water cage or vat directly. The weight of the active substance as a unit of measurement typically does not take into account differences in potency and bioavailability for the various drug substances (Grave et al., 2004).

One of the oldest areas of statistical application is the modeling of safe levels of exposure to chemicals, treatments and pharmaceutials products. Toxicology and much dose-response methodology is based on use of the logistic curve and the idea of a threshold affecting useful intake of treatment dosage. This area has generated a large number of useful nonlinear regression models and summary measures such as median survival level (LD50) are commonly used (Morgan, 1992). More recently survival analysis based proportional hazards or logistic regression models have been developed. Often the ROC (receiver operated characteristics) curve can be generated in such models, a combination of the sensitivity and specificity of the underlying treatment or assessment informative as a diagnostic.

Clinical trials are the more formal approach to take when developing drugs and their appropriate dosages for human or other species, and there is a formal phase I - IV developmental path that is followed. This is not examined in detail here with each phase involving different experimental settings and related statistical methods.

While treating existing infections is important and often based on antibiotics, the issue of how to best vaccinate to give immunity to an infectious agent in a population is also an old one. In human populations the use of weak or dead infectious agents as the basis of vaccine, training the immune system, for example in regard to cowpox infection as a natural guard to smallpox infection, or administering polio vaccine. Similar tactics can be used in animal populations. In fish populations and other aquatic populations temperature may play a role and can allow the immune system time to adjust to the presence of an infectious agent such as sea lice.

10.3 Data

The data here is simulated, based on a study of sea lice infestation and treatment in a rainbow trout population (Beaman et al., 1999, 2001). It is known from this previous work that temperature alone can lower the level of infection, allowing for improved development of immune response in populations of rainbow trout. We now seek to augment this effect by adding a dose response effect using an antibiotic administered through the feed placed in each tank.

The treatments (temperature level of initial sea lice exposure (2 levels) and subsequent antibiotic dosage (3 levels)) are randomly assigned by tank. There are three dosage levels applied at random to 3 tanks within each temperature level, each with 20 fish. A single dose is given for each tank and the fish are each inspected and weighed a 5-day period. Water temperature for each tank is carefully controlled and initially set with low (7 degrees C) or typical water temperature (15 degrees C). The antibiotic is added to the feed after sea lice infection. There is no way to assess directly the biochemical intake of the antibiotic in each individual fish.

Each tank is given a specific randomized antibiotic dosage, three after being subjected to colder water temperature when initially infected with sea lice, three after being infected with sea lice at regular temperature. A control group is also maintained for comparison purposes. The results for all groups are examined over the same basic time period. The data variables are given in Table 10.1.

10.4 Specific Aims, Hypotheses, Models

The improvement in sea lice infection level is the primary outcome of interest. This is scored on a continuous scale (0 - 10). This is to be related to the level of dosage and water temperature of first exposure. Other health related

TABLE 10.1
Variable Listing

Dosage level	0, 10, 20
Water Temperature	7, 15
Weight	Continuous
Length	Continuous
Sea Lice Graded	Continuous 1 - 10
Sea Lice	Hi/Low
Gene A	Yes/No

outcomes such as weight and length are also of interest and may also be used as control variables. It is thought that the presence of a specific gene may povide some protection from sea lice infection. This is assumed to be randomly spread through the entire set fish being studied.

Two-way ANOVA is the standard model for analysis here. It can actually be seen as a 1-way ANOVA with a blocking variable (temperature) given the randomization employed. Homogeneity of variance is not necessarily the case, and should be examined. If it is present a transformation of the measurement scale may be necessary or use of weighted least squares. No attempt is made to develop a detailed pharmaceutical based analysis.

Prior densities to be taken in regard to the parameters in the various models here can be non-informative or simply Normal with diffuse variation. Here we use improper priors though the chosen priors will differ depending on the model and likelihood applied. ANOVA and ANCOVA are the main contexts here.

The basic questions of interest here are:

1. Is there a difference between the dosage level in relation to controlling the onset of sea lice? Use ANOVA testing.

2. Does the temperature at which sea lice exposure is initiated affect the usefulness of the antibiotic?

3. Can we predict the level of sea lice infection at specific dosage and temperature levels?

10.5   Analysis and Interpretation

10.5.1   Data Analysis

The data is simulated based on the studies in Beaman et al., (1999), (2001). The data is available in the supplementary materials. The analysis is based

TABLE 10.2
Data

	Dosage = 0	Dosage = 10	Dosage = 20
Temperature = 7 (n, mean, std dev)	(20, 41.20, 1.824)	(20, 24.7, 2.029)	(20, 16.85, 3.52)
Temperature = 20 (n, mean, std dev)	(20, 63.85, 3.031)	(20, 39.25, 2.197)	(20, 23.45, 2.06)

on a basic two-way ANOVA format and design. Basic data summary values are given in Table 10.2.

Scatterplots of sea lice index versus dosage and temperature are shown:

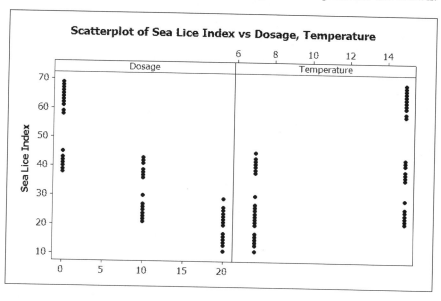

FIGURE 10.1
Scatterplots of Sea Lice Index versus Dosage and Temperature

10.5.2 Likelihood Function

Assuming a normal error distribution here as well as a linear regression structure we have as our initial overall likelihood function:

$$L(\boldsymbol{\beta}, \sigma^2 | \mathbf{y}) = c \cdot (1/2\pi\sigma^2)^{n/2} \exp(-(1/2\sigma^2) \sum_i (y_i - \beta_0 - \beta_1 x_{1i} - \cdots - \beta_p x_{ip})^2)$$

The formal hypotheses in question here are in relation to the β_j parameters

TABLE 10.3
ANOVA Table

Variable	Source			
	df	SS	MS	p-value
Dosage	2	21470.3	10735.2	0.00001
Temperature	1	6394.8	6394.8	0.00001
Error	116	2013.2	17.4	
Total	119	29878.4		
R^2		.668		

and the analysis from the frequentist perspective is given by the standard ANOVA table and related least squares based tests and confidence intervals.

If we have a cutoff regarding an acceptable level of sea lice infection, then this threshold can be used to dichotomize the response and potentially employ a binomial error distribution here and a logistic regression structure. If the data supports it, we will have as our initial overall (generalized linear model based) likelihood function:

$$L(\boldsymbol{\theta} \mid \mathbf{y}) = c \cdot \prod_{i=1}^{n} \left[\frac{\exp(\sum_{j=0}^{k} \theta_j x_{ij})}{1 + \exp(\sum_{j=0}^{k} \theta_j x_{ij})} \right]^{y_i} \left[1 - \frac{\exp(\sum_{j=0}^{k} \theta_j x_{ij})}{1 + \exp(\sum_{j=0}^{k} \theta_j x_{ij})} \right]^{1-y_i}$$

This model can be used to obtain the ROC curve, odds ratio estimates and extend the analysis. Note further that if this experiment were extended to include multiple replications over time, or season, the infection rate could be examined using a Poisson distribution.

10.5.3 Likelihood Frequentist Analysis

The linear model (ANOVA) maximum likelihood model parameters are estimated using least squares and standard errors estimated from the appropriate variance-covariance matrix. P-values reported for hypotheses in question use standard pivotal quantities based also on the *m.l.e.* which in the linear model setting with normal error are the standard t or F based statistics.

Using the two-way ANOVA with no interaction (additive model) we have our ANOVA based testing (Table 10.3).

Residual plots support the choice of likelihood in the sense that the pattern in the residual plot does not contradict the assumed likelihood.

Note that dosage is applied to tank randomly here within each Temperature setting. So the experiment is not completely randomized. In a sense it is a simple block design with each Temperature level serving as a replicate or block for the experiment. In this type of setting interaction terms are typically not utilized.

We can also view each Temperature level as two distinct blocks here and

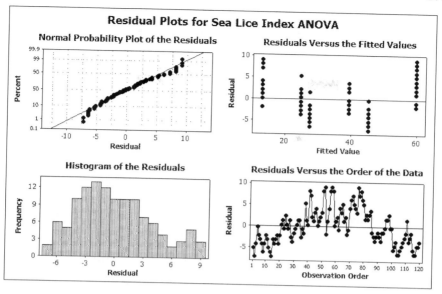

FIGURE 10.2
Residual Plots

examine dosage effect for each level; this gives significance at both levels. Dunnett's (Bonferroni type) correction for comparing dosage levels to the control level (0 units) also gives a significant difference for both 10 unit and 20 unit levels. In both temperature levels the difference increases as the dosage rises. This is also apparent from the data plots.

10.5.4 Likelihood Bayesian Analysis

ANOVA based testing from the frequentist perspective reflects in large part the presence and use of randomization in the experimental process. Technically this is not required from the Bayes perspective, though is obviously relevant to the generation of an unbiased sample and experiment. Indeed the ANOVA table itself arises in regard to specifically chosen priors.

For example, if noninformative priors are used for the β_j and σ^2 parameters, then essentially the joint posterior density is the likelihood function. Thus if the likelihood itself is Normal as above, the posterior density of the vector of parameters β is the multivariate-t distribution. If this holds, then the posterior density of the quantity $Q(\beta) = (\beta - \widehat{\beta})'X'X(\beta - \widehat{\beta})$ has an F distribution a-posteriori and in a sense justifies the quadratic form based ANOVA testing format from the Bayesian perspective. The results here are therefore relevant from the Bayesian perspective if non-informative priors are assumed.

There are however many approaches, especially if the use of improper priors is rejected or non-Normal likelihoods are to be used. This is especially true as MCMC based computational approaches have broadened the use of a variety of hierarchical models. See for an early reference Lindley and Smith (1972). This can lead to slightly different estimates and results as the question of which standard error to use for interpretations arises and which prior is chosen.

Given the close relationship between frequentist and Bayesian results in the setting of a normal likelihood, here we employ the Bayesian perspective to investigate the predictive distribution of sea lice in regard to a specific dosage (20 units) within each of the temperature levels employed, comparing the distributions. This allows for careful use of the Bayesian predictive distribution as an added component of the overall analysis (See Suggested Exercises). In the ANOVA setting where randomization plays a clear role in defining the ANOVA model of interest it seems good statistical practice to accept that the frequentist result has a clear Bayesian interpretation. A distinct Bayesian analysis can be applied here in regard to the predictive distribution or in the case of existing theory that supports specific prior density and prior belief to be incorporated into the analysis.

10.5.5 Analysis/Integration/Comparisons

As we move from data analysis to assumed likelihood and likelihood based inference to the Bayesian perspective on the likelihood function and Bayesian inferential concepts, in the ANOVA setting the issue of to what extent considerations such as randomization and family-wide Bonferroni corrections should be included in statistical summaries of experiments arises.

The frequentist analysis is based on the properties of the likelihood, which in the case of normality is essentially least squares, guided by the othogonal decompositions that underlie the ANOVA testing format. Here the structure of the experiment is an additive two-way ANOVA which can also be seen as a 1-way with blocking on the temperature variable. The temperature and dosage variables are seen to be significant. Lower temperature again supports a lower infection load (lower number of sea lice per fish) and higher dosage level lowers the infection load at both higher and lower temperatures.

The Bayesian analysis here is reflected in the ANOVA table to some extent assuming non-informative priors. Other priors or likelihood choices (in smaller samples) would lead to a different set of answers. That depends on the choice of the researcher. A more detailed Bayesian analysis would also employ a hierarchical model which is not covered here. It is worth noting that likelihoods and priors need to be carefully chosen in such settings. Not all prior-likelihood combinations will yield well-defined posterior densities.

10.6   Suggested Exercises

1. Consider transformations for homogeneity prior to running the ANOVA anlyses above. Some approaches include the Box-Cox method and variance stabilizing transformation. Note that Bayesian incorporation of the Box-Cox adjustment implies searching for a prior for the parameter in question. Data transformations simply rescale the observed data from the Bayesian perspective and thus are acceptable.

2. Use other more robust distributions to derive the likelihood here. Maintain the selected priors if the parameters have a similar interpretation. Repeat the basic analysis and quantify any observed differences.

3. Develop a hierarchical model approach to the additive two-way ANOVA. If we assume potential outliers, a likelihood based on the Student-t distribution can be assumed. Re-analyze the data using this likelihood and several prior densities (use proper priors).

4. Develop a predictive density for sea lice index measure for the entire dataset. Use a Normal prior and assumed independent normally distributed future observation. Obtain a predictive sea lice density for each initial temperature and dosage subgroup. Plot them and discuss any differences.

5. Use Bayes factors to compare various models for sea lice infection, where we choose among the set of explanatory variables above. Do we see a clear pattern in the Bayes factor values as we add information (each variable)?

6. Derive the predictive distribution for sea lice infection for both dosage $= 20$ and dosage $= 0$ when temperature $= 15$. Compare the two densities visually and also using the Kullback-Liebler distance. Do they contain information regarding the effect of dosage on sea lice infection? Could we compare all 6 sea lice infection groups in this manner?

10.7   Bibliography

[1] Beaman H.J., Speare D.J., Brimacombe M. (1999). Regulatory Effects of Water Temperature on Lomae Salmonae (Microspora) Development in Rainbow Trout. *Journal of Aquatic Animal Health* 11, p. 237–245. DOI:10.1577/1548-8667(1999)011<0237:REOWTO>2.0.CO;2.

[2] Beaman H.J., Speare D.J., Brimacombe M., Daley J. (2001). Evaluating Protection Against Loma Salmonae Generated from Primary Exposure of Rainbow Trout, Oncorhynchus mykiss (Walbaum), Outside of the Xenoma Expression Temperature Boundaries. *Journal of Fish Diseases*, 12/2001; 22(6):445–450. DOI:10.1046/j.1365-2761.1999.00194.x.

[3] Grave K., Horsberg T.E., Lunestad B.T., Litleskare I. (2004). Consumption of Drugs for Sea Lice Infestations in Norwegian Fish Farms: Methods for Assessment of Treatment Patterns and Treatment Rate. *Dis Aquat Org,* Vol. 60: p. 123–131.

[4] Lindley D.V. and Smith A.F.M. (1972). Bayes Estimates for the Linear Model. *J. Roy. Statist. Soc. B. 34*, Vol 1, p. 1–41.

11

Patterns of Genetic Expression in Mouse Liver Cancer

11.1 Science

Genetics is a major area of application of science and statistics. With the advent of genomics and much more detailed genetic information, there has been a renaissance of statistical applications in genetics. A major area of interest is the understanding of genetics in regards to cancer and its treatment. The uncontrolled growth of cells characterizes cancers and typically implies a change in healthy cells, especially on a genetic and cellular level.

Patterns in gene expression that are related to the onset of cancer are an area of great research interest. Some studies have a small number of subjects and a relatively large number of measured variables. The application of linear models and related PCA based techniques in such settings are referred to as high dimensional data analysis. This is a new research area that is challenging, requiring modification and careful application of standard statistical models and related concepts (Schadt et al., 2005).

The models that underlie cancer, defined here as highly unrestricted cell growth, tend to be mathematical in nature, often reflecting cell counts and the number of cancer cells accumulated over time. This may also be augmented with specific measures of cell morbidity or deformity. Once such cells become highly numerous the pathology of the organ infected alter form and function. This may also be an element of the study. There are often threshold levels associated with the study of cancer cells, especially when treatment has been introduced and the cells involved are controlled, the cancer passing into a defined remission.

The causes of cancer, while highly associated with family history, have recently been identified as primarily random in nature, occurring at the level of stem cells approximately $2/3$ of the time. Often the generation of such cells is then met with the activity of the immune system, controlling the practical effect of cancer for many individuals. This points to the effect of age in relation to cancer. As the immune system weakens due to age, the presence of cancer is higher. The immune system functions as a control on the number of cancer cells and non-standard cells in general in the system. Many other network type effects have been observed in relation to cell growth and cell and function regulation. When these controls break down, the spread of cancer may occur.

In the simulated example here, microarray genetic marker data from chromosome 11 in an F2 mouse intercross is used to examine large-scale gene co-expression in the liver and its relation to a cancer severity index. We then attempt to also define how the expressed genes correlate with each other. Much work has recently gone into developing potential networks or sets of correlated genes that as a group affect the onset of various phylogenetic outcomes (Ghazalpour et al., 2006; Tanizawa et al., 2010).

While genomics has become available and the computing technology and methods to incorporate such large amounts of data, most genomic data in a genome-wide analysis will not be directly relevant to the analysis at hand and indeed may clutter the analysis with spurious or irrelevant correlations. Here a fairly simple simulation is conducted based on a mouse genetic data experiment and analyzed to give an overview of some of the issues that arise.

The challenges in this type of work should not be underestimated and they often are. When not guided by a serious scientific model, here in relation to the function of the genes themselves, individually and as a cluster, there is a heavy dependence on the statistical assessments involved. These can be affected by weaknesses in the design of the study and resulting data which may simply be the outputs of various Affymetrix chips and/or bioassays, quilted together from various public databases to form an impressively large dataset. Statistical methods in such settings may be of little value, indeed can easily fall prey to bias, heterogeneity and non-standardized scaling (Lambert and Black, 2012). Replication of the results is rarely a component of the study design.

Population genetics considerations can also arise in the genomic study of genetic mutations. The models of population genetics model the flow of genes through the generations, invoking such concepts as dominance, linkage, recessive genes etc., all present in genomics which often reflects the current generation as we are each the end product of many generations. Gene mutations that have primarily negative effects on the fitness of the organism or phenotype in question often disappear quickly from the population under study. Genome-wide studies will reflect potentially high levels of such irrelevant mutations, and this may affect statistical inferences.

In the analysis of large genomic expression databases the Bayesian perspective serves as a way of averaging and stabilizing calculations in high dimensional models, with the use of technically useful convenience prior densities dominating (Park and Casella, 2008; Mallick and Yi, 2014). There are often few theoretical expectations in new areas of genomic analysis as much of this remains to be developed. Data analytic approaches are very common and such computer algorithmic search must be carefully interpreted when the algorithm is based on linear models and essentially linear concepts such as correlation. Mis-specification of a linear model when the underlying patterns are nonlinear

TABLE 11.1

Variable Listing

Gene j Expressions	Continuous
Immune cell count	Continuous
Oxidative stress	Continuous
Overall cell growth	Continuous, Percentage
Cancer severity index	Continuous
Categories re cancer development	Discrete

can lead to serious issues in the stability and relevance of the fitted model, especially in higher dimensional settings (Brimacombe, 2017).

Large databases in this context often have many gene expression variables and relatively few subjects ($p > n$), so called high dimensional databases. These cannot be analyzed from the frequentist perspective as $n > p$ is required for an identifiable likelihood where *m.l.e.* values can be obtained. Typically a restriction is placed on the model, for example a sparsity restriction that only allows a given number of genes to be considered at one time and cycles through all possible sets of genes to obtain a best fitting model (Efron et al., 2004).

Finally in any model where there are large numbers of variables and therefore parameters, many hypothesis tests are run on the same basic dataset and model, giving rise to multiple comparison issues. These affect frequentist calculations where they are addressed with False Discovery Rate assessment and Bonferroni corrections. In the Bayesian setting the very high number of integrations required for inference can be viewed as creating a shrinkage or overfittting effect and all results need to be carefully interpreted.

11.2   Data

The data is simulated based on a study of the genetics of mouse livers (Ghazalpour et al., 2006). The specific example developed here focuses on Chromosome 11 only. Broadening the dataset here to include more chromosomes does not alter the basic approach, only the complexities involved. The data collected include the expressed genes, the cancer severity index and also immune cell count, oxidative stress and a categorized measure of cancer severity (Low, Medium, High). A brief data dictionary is given in Table 11.1.

11.3   Specific Aims, Hypotheses, Models

This type of analysis is heavily data oriented so there is less focus on mathematical models beyond simple linear and sparse linear models, if these deserve to be called models. There is also relatively little scientifically validated belief in regard to scale (linear, non-linear), correlation structure and the specific relationship between the expression of Gene j and the cancer severity measure. Given the limited scientific structure currently available for most genetic settings, the overall strategy here is to try to observe and understand data based patterns in mice with liver cancer. Much analysis of this type is most easily considered as data and cluster analysis, sometimes referred to as hypothesis generating. Such results or data patterns require careful replication if possible.

The primary outcome of interest is a cancer tumor (size, weight and stage based) severity index. This is taken to be continuous (1, 100). In relation to this we have a set of 20 genes restricted to Chromosome 11 here. In terms of models, linear models are typically employed as a default. Most analysis reflects Singular Value Decomposition (SVD) here, applying PCA based cluster analysis for the continuous gene expression data. These reflect the eigenvalues of the underlying matrices. Further classification trees are employed to examine the clustering of both genes and also individuals. Generating such empirical classification structures is possible for groups and this can be done here for cancer severity high and low groups (using the median value as a threshold).

In terms of likelihood based analysis, the normal distribution is assumed here and the *m.l.e.* are least squares estimators and standard linear regression analysis can be conducted. As well, a linear scale is assumed and we do not alter the scale here. More detailed analysis might look to rescale the expression data, both in terms of standardization and also on a logarithmic scale.

Bayesian analysis can apply the normal distribution along with fairly disperse prior densities. The large samples involved also tend to move the resulting joint posterior and marginal posteriors to normal distribution based results. If past cluster based analyses are available, the assumed prior densities can be given a normal shape along with more correlated structures among the beliefs regarding the parameters in the linear model. Marginal posterior densities give credible regions for parameters.

Note that in the case of $p > n$ high dimensional models, the use of priors results in a posterior, very much an augmented likelihood function here, that is identifiable. Thus the prior density takes on a greater role than simply representing existing beliefs regarding the parameters in the linear model linking cancer severity and gene expressions. Further, if a sparsity restriction is included in the prior density chosen, this is empirical Bayes in a fundamental way as we are using the observed data structure, $p > n$, to guide the formation of the prior density.

Here the basic questions of interest are:

1. Can a linear model relating the cancer severity index be developed here?

2. Can a linear model reflecting the Bayesian context be fit to the data?

3. Are there correlation based cluster patterns in the data, among the gene expression data?

11.4   Analysis and Interpretation

11.4.1   Data Analysis

The data is simulated based on Ghazalpour et al., (2006). The example developed here focuses on Chromosome 11 and has a much smaller number of variables and individuals. A cancer severity index measure is derived and gene expression for 56 genes obtained on 116 individuals. The goal is to (i) relate the genes to the cancer severity measure and (ii) obtain existing clusters, if any, among the genes. The actual scale of the numbers in such studies tends to have litle absolute meaning, reflecting the need to scale the values to improve comparability across bioassays.

Examining large amounts of data and variables is a challenge and the number of cluster related methods has grown quickly over the past years. Not all datasets possess useful patterns and analysts need to be careful not to impute structure where there may be none. Here we focus on standard approaches to large multivariate datasets that can be interpreted and whose limitations are known.

The gene expression values are rescaled and standardized to have a $(0, 100)$ index value. The overall cancer severity measure has been similarly scaled. A dendrogram for the set of gene variables is available. No clear pattern can be detected.

Prinicipal components analysis is often useful to begin the study of such data. A loading plot showing the loading of the variables (the expressed genes in this case) on the principal components is obtained.

Of interest here is determining the number of independent sources of variation in the overall dataset. Here this is done through the principal components themselves and are by definition uncorrelated with each other. If the original dataset if approximately normal then the principal components can be interpreted loosely as independent. The number of important components can be visualized in the scree plot:

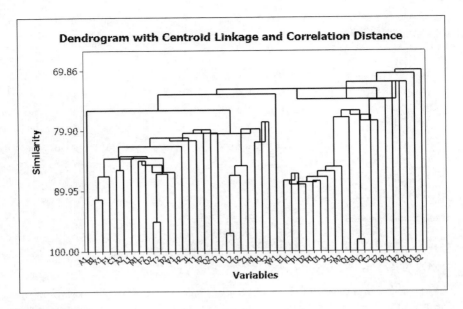

FIGURE 11.1
Dendrogram Plot

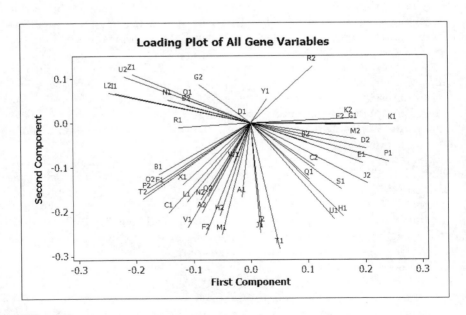

FIGURE 11.2
Loading Plot

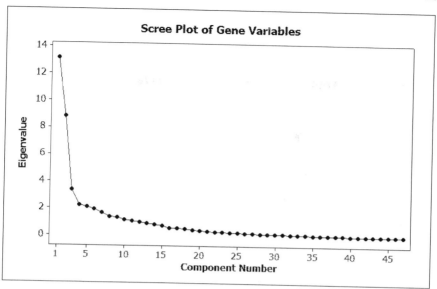

FIGURE 11.3
Scree Plot

The number of principal components required to explain 80% of the total variation in the data is approximately 11 (80.8%). This is where the scree plot graphic is approximately one.

Another use of principal components is to plot the first two principal components against each other. These are the two PCAs that explain the largest amount of variation on their own. Typically if there are outliers or non-standard clustering patterns this can be a way find them. Here the plot shows no outlier values.

11.4.2   Likelihood Function

Assuming a normal error distribution here as well as a linear regression structure we have as our initial overall likelihood function:

$$L(\boldsymbol{\beta}, \sigma^2 | \mathbf{y}) = c \cdot (1/2\pi\sigma^2)^{n/2} \exp(-(1/2\sigma^2) \sum_i (y_i - \beta_0 - \beta_1 x_{1i} - \cdots - \beta_p x_{ip})^2)$$

The hypotheses in question here are based on the β_j values. Note that there may be no actual justification for assuming genomic expression follows a normal distribution, however the large sample asymptotic nature of the dataset implies results that will converge to those given by assuming a normal density.

Another approach is to attempt to restrict the data to cases of high cancer

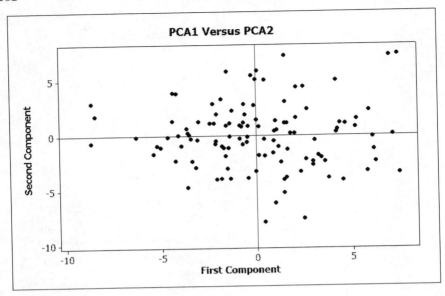

FIGURE 11.4
Plot of First Two Principal Components

severity, arguing that it is in these cases that the genes will have expressed most strongly, though we lose the relative comparison with lower severity cases. A principal components analysis here shows that 10 components are required to explain 81.3% of the total variation in the data.

11.4.3   Likelihood Frequentist Analysis

The use of data analytic methods tends to fall under the frequentist label, but does not reflect frequentist probability assumptions. In settings when the scientific insight is limited and data analytic approaches do not detect basic evidence of clustering patterns, it is difficult to argue for patterns detected by the assumptions of (i) a mathematical model and likelihood function and (ii) prior densities for parameters defined within the model. The data analysis presented above has assumptions regarding underlying linearity in the relationships between variables not in regard to a potential likelihood or prior density.

If a linear model is employed then we obtain the model:

Cancer severity = 74.6 + 25.9 A1 - 175 B1 + 83.7 C1 - 61.2 D1 - 81.4 E1
+ 17.2 F1 - 222 G1 + 51.9 H1 - 170 I1 + 88.3 J1 - 180 K1 + 25.7 L1
- 8.4 M1 + 0.8 N1 - 56.2 O1 - 59 P1 - 35.5 Q1 - 37.0 R1 - 88.5 S1
- 123 T1 + 31.5 U1 + 15.7 V1 + 66.0 W1 + 30.3 X1 + 32.3 Y1 - 56.6 Z1
- 120 A2 + 93.0 B2 - 2.1 C2 + 21.2 D2 - 34.8 E2 + 11 F2 + 15.0 G2
+ 21.5 H2 + 39.2 I2 + 35.5 J2 + 206 K2 + 135 M2 + 210 L2 - 3.9 N2
+ 26 O2 - 9.2 P2 - 51.4 Q2 - 172 R2 - 5.5 S2 - 226 T2 - 18 U2
$R^2 = 49.5\%$

This is not a particularly useful model.

Principal components regression is also a possibility, though this can lead to difficulties and misleading results. Another option is the use of a sparsity restriction as mentioned in the text below. Here the linear model (and thus the likelihood function) is restricted to a stagewise approach and only a few variables (in this case 10 or 11 would be a useful initial guess) are analyzed at a time. After a variable is chosen the residual aspect of the response vector is then modeled with the remaining variables, choosing the best fitting variable according to the C_p criteria. See Efron et al., (2004). This continues until the selected number of variables in the sparsity condition is attained. There are many versions of this, including Bayesian versions using other criteria. Standard stepwise regression here provides a model of 10 gene variables giving an R^2 of 19%.

If the number of variables is greater than the number of subjects, the likelihood is not identifiable in regards to determining its maximum. One way around this is to assume a sparsity restriction on the model fitting process. This assumes that only a small number of the samples variables are actually related to the outcome in question.

A sparseness restriction can be expressed:

$$\sum_{i=1}^{k} |\beta_i|^m < t,$$

for relatively small chosen values t and k. Sparseness restrictions typically assume $m = 1$ or 2. This is most useful when only a few x_i are significantly correlated with the response y_i. It is also necessary since in these settings the usual least squares estimator $\hat{\beta} = (X'X)^{-1}X'y$ will not exist. Interestingly, from the perspective of the parameter space, such restrictions limit the set of possible β_i combinations that may be examined. The shape of the restricted parameter space depends typically on the value of m.

The LASSO approach using the lars algorithm can be applied here. Sparse model restrictions can be seen here as extended use of stepwise and ridge regression procedures. It is fair to say they also inherit the criticisms that affect stepwise procedures, such as the use of many sequential p-values and tests to find the optimal model without necessarily correcting for these steps. The lars algorithm uses the geometry of least squares and orthogonal projection to avoid some of this criticism and is fairly stable.

Note that just as with the initial data analytic approaches, most of these higher dimensional models, from a frequentist or Bayesian likelihood perspective, assume basic linearity in gene expression on the employed scale. This is somewhat questionable, especially as recent advances have begun to link aspects of gene expression and information to the physical three-dimensional structure of the chromosome. This implies many more possible linkages in a nonlinear fashion.

11.4.4   Likelihood Bayesian Analysis

The use of a linear model from the Bayesian perspective does not improve the performance of the linear model here. It is useful to consider possible priors for the specific set of genes considered here. There is some existing theory to guide the application of priors, typically in the form of a normal distribution with diffuse variation (non-informative) and chosen mean values. If desired the mean values here can be chosen to reflect an expected level of correlation with the cancer severity measure. However there is little previously known structure in the data to guide prior selection (this does not stop much creative activity in this direction) and assuming a likelihood based on the normal distribution or a prior density that is non-informative, either improper or a symmetric density with large variation. seems reasonable.

Under a joint prior with independent normal assumptions for each parameter ($\beta_j \sim N(0, .0001)$), improper flat priors and an inverse gamma density for σ^2 throughout, we have for the first 5 listed genes (Table 11.2).

11.4.5   Analysis/Integration/Comparisons

Genetics is challenging, often reflecting structures that are often not incorporated into genomic analyses. Phenotypes follow distributions therefore the genomic structures underlying phenotypes may also follow distributions making it difficult to see patterns when examining a single datset or sample. As well gene expression may be linked to growth or non-constant, even nonlinear patterns of expression in relation to maintenance of existing genetic network or other expression patterns. Linear and correlation based models are necessary to deal with the very large amounts of data, but may not reflect actual genetic structures.

The issue of the number of variables in genomic studies reflects the lack of understanding of how many genes function and relate to the functioning of proteins in the development of specialized cells. Such a large amount of information would seem to require more subtle and complex structures that can modify over time or in relation to challenges. The presence of gene expression may lack appropriate context. Note that the lack of replication of many genomic results reflects this difficulty in interpretation and without replication the scientific support for a statistical result is limited at best.

As we move from data analysis to assumed likelihood and likelihood based

TABLE 11.2
Ninety Five Percent Confidence and Credible Intervals

Prior Assumption	A1	B1	C1	D1	E1
Non-Informative	(−15.9, 64.2)	(−314.3, −31.2)	(−38.1, 201.6)	(−143.5, 31.1)	(−242.9, 93.3)
Normal	(−14.7, 62.1)	(−303.8, −34.9)	(−34.7, 197.3)	(−138.9, 29.7)	(−237.2, 90.7)
m.l.e. No prior	(−16.2, 67.9)	(−321.4, −28.9)	(−40.0, 207.4)	(−154.7, 32.3)	(−258.2, 95.4)

inference to the Bayesian perspective on the likelihood function and other Bayesian inferential concepts, we are slowly making more mathematical and structural assumptions regarding the scientific problem itself. This must be done carefully in the genetics setting where the function of genes remains unknown in most cases. With only 20,000 genes, not the 100,000 that were expected early on, it is obvious that in order to protect the passing of information from generation to generation, genomic information is based on moderately few elements, many of which are re-used in many different ways. This research is ongoing and as the science finds firmer ground, the statistical methods required will alter, perhaps re-focusing on fewer genes in relation to a particular outcome.

The Bayesian prior here often does not reflect past studies. The computational burden is often onerous and priors are often chosen to allow for efficient application of MCMC based numerical integration procedures. This is somewhat anti-Bayesian as choosing priors due to the form and integrability of the resulting posterior density does not reflect prior knowledge, but rather properties of the observed likelihood for the current analysis. This should be approached with caution if the analysis is to have a formal Bayesian interpretation. Without a clearly defined prior baseline of knowledge, how is a learning model to be interpreted?

11.5   Suggested Exercises

1. Standardization is often an important first step in the analysis of genetic data, especially when there is often more than one bioassay involved and this generated unwanted non-constant variation. It often rescales the data for the sake of comparability. What is the effect of such rescaling on the priors to be used, especially as some rescaling may alter the correlation among the variables in the model.

2. The use of large samples implies the presence of asymptotic results. Discuss the expected similarity of Bayes and frequentist results in the setting of linear models when we can argue that standard regularity conditions hold.

3. Examine the use of less precise priors, for example the student-t distribution. How does this affect the accuracy of the regression model in such large samples?

4. In all genomics settings reflecting only the current generation. Discuss whether there is a need to ensure that the observed data has specific population genetic aspects or structures. For example mutations should arise and there should be quite a few that impact

negatively on the fitness of the organism. These will disappear in the next generation so they are not relevant.

5. Discuss issues of $p > n$ prior selection when $p > n$ and the likelihood function lacks identifiability. What are the possibilities? How can the assumption of sparsity as an element of the prior density be justified?

6. Using available gene network software, from both Bayesian and frequentist perspectives, represent the existing correlation structures in the data as a gene network.

7. The issue of mis-specification can often lead to incorrect inferences when for example linear models are used when nonlinear relationships exist among the variables in a model. Discuss how higher order effects may not be detected by simpler linear models.

11.6 Bibliography

[1] Efron B., Hastie T., Johnstone I. and Tibshirani R. (2004). Least Angle Regression. *Annals of Statistics*, 32, No. 2, p. 407–451.

[2] Schadt E.E., Lamb J., Yang X., Zhu J., Edwards S., GuhaThakurta D., Sieberts S.K., Monks S., Reitman M., Zhang C., Lum P.Y., Leonardson A., Thieringer R., Metzger J.M., Yang L., Castle J., Zhu H., Kash S.F., Drake T.A., Sachs A. and Lusis A.J. (2005). An Integrative Genomics Approach to Infer Causal Associations Between Gene Expression and Disease. *Nat. Genet.* Vol 37, No. 7, July 2005, p. 710–717.

[3] Ghazalpour A., Doss S., Zhang B., Wang S., Plaisier C., et al., (2006). Integrating Genetic and Network Analysis to Characterize Genes Related to Mouse Weight. *PLoS Genet* 2(8): e130. DOI: 10.1371/journal.pgen.0020130.

[4] Lambert C.G. and Black L.J. (2012). Learning From Our GWAS Mistakes: From Experimental Design to Scientific Method, Biostatistics 13, 2, p. 195–203.

[5] Mallick H. and Yi N. (2014). Bayesian Methods for High Dimensional Linear Models. *J Biom Biostat.*;1: 005–. doi:10.4172/2155-6180.S1-005. p. 1–27.

[6] Morgan B.M.J. (1992). *Analysis of Quantal Response Data.* Chapman & Hall, New York.

[7] Park T. and Casella G. (2008). The Bayesian Lasso, *J. Am. Statist. Assoc.* 103, Vol. 482, p. 681–686.

[8] Tanizawa H., Iwasaki O., Tanaka A., Capizzi J.R., Wickramasinghe P., Lee M., Fu Z., and Noma K.-I. (2010). Mapping of Long-Range Associations Throughout the Fission Yeast Genome Reveals Genome Organization Linked to Transcriptional Regulation. *Nucleic Acids Research*, 38, No.22, p. 8164–8177.

12

Antibiotic Resistance in Relation to Genetic
Patterns in Tuberculosis

12.1 Science

The ability of disease pathogens to resist treatment is an area of research in
which evolutionary considerations and genetics interact. Statistical modeling
and the design of the particular study of interest are key components in this
area of research. Tuberculosis is a very old scourge of humanity, dating back
thousands of years. It tends to return in settings of hardship, poverty and
related diseases. As with many older infectious elements, its relationship with
patients has evolved and infection does not immediately kill the patient, but
slowly hinder the life of the infected. Treatment is time consuming and antibiotic use is a key component. Recently antibiotic resistant strains of TB have
become a serious issue (Cohn et. al, 1997).

In a world where international travel is common, the need for global surveillance of many infectious diseases is a real one. The variability in the case of
TB of resistant cases requires modeling and perhaps detailed genetic analysis to better focus treatments and awareness of possible resistance. Several
studies have looked at the genetics of M. tuberculosis itself in relation to
various types of antibiotic resistance (Hazbon et al., 2005, 2006). Parasites,
bacteria and viruses tend to have relatively shorter genomes and these are
more amenable to analysis with regards to a specific outcome, in this case
specific types of antibiotic resistance. This type of analysis is likely to have
been underpowered with the smaller data sets previously available.

More recently TB surged in areas of the old Soviet Union in relation to
the onset of AIDS and related immuno-suppressant related illnesses. In Africa
and areas of India, TB remains a common illness. Resistant TB is a serious
worry and the study of resistance of TB to standard antibiotic treatment
is important. TB reflects a clade or SGC group based categorization and
genetically reflects greater variation in areas of the world where TB strains are
older. This heterogeneity may arise in gene expression studies (Brimacombe
et al., 2007).

INH is the basis of antibiotic treatment for drug-susceptible tuberculosis
and often used to treat latent M. tuberculosis infection. Recent increases in
INH-resistant and multidrug-resistant (MDR) tuberculosis are now limiting

the usefulness of this drug. Often INH resistance is a first step in the development of MDR. Therefore there is serious interest in identifying genes associated with INH resistance in clinical M. tuberculosis isolates.

In regard to specific mutations in the TB parasite that might confer resistance, a recent study (Hazbon et al., 2006) showed that between 40 and 95% of INHr clinical M. tuberculosis isolates had mutations in katG315 which may be favored as mutations here appear to decrease INH activation without limiting catalase-peroxidase activity, a virulence factor. M. tuberculosis may compensate for katG mutations by overexpressing the ahpC gene. INH resistance can also develop through alterations in the INH drug target InhA. Mutations in ndh, a gene encoding an NADH dehydrogenase confers resistance to INH and ethionamide in M. bovis. Mutations in at least 16 other genes have been associated with INH resistance in clinical isolates, however, the roles of these genes in INH resistance remain unclear.

Antibiotic resistance of TB is a worldwide problem but there are limitations to its assessment in developing countries having few facilities. If available, standardized laboratory methodologies and bioassay development may not be followed. Work in Cohn et al., (1997) reviewed 63 surveys of resistance to antituberculous drugs performed between 1985 and 1994. The study reported rates of resistance to commonly prescribed antibiotics in TB cases, noting practical limitations. Rates of multidrug-resistant tuberculosis were: Nepal (48.0%), Gujarat, India (33.8%), New York City, (30.1%), Bolivia (15.3%), and Korea (14.5%). In South Africa, the acquired rate was 10.9%, and primary resistance rate 3.8%. In Korea, the United States, and Argentina, the rate of acquired resistance versus primary resistance were 41.9% vs. 12.6%, 10.4% vs. 3.2%, and 4.2% vs. 1.6%. In Libya, the rate of acquired resistance vs. primary resistance to streptomycin was 6.8% vs. 2.9%. Selection of patient samples for susceptibility testing was variable between and within countries and often the sampling was not well described.

In Hazbon et al., (2005), (2006) an international set of collected TB samples was analyzed using in part a logistic regression model which was developed with predictive concordance of over 80% based on a specific set of genes that were interpretable in relation to TB and resistance. The simulated dataset is based on these studies.

12.2   Data

The data here are a simulation based on an earlier study (Hazbon et al., 2006) of antibiotic resistant TB drawn from an international set of samples. This analysis is replicated at two distinct time periods with samples drawn from distinct samples with some minor overlap. The variation may be affected by bioassay and other measurement related effects (Table 12.1).

TABLE 12.1

Variable Listing

INH Resistant	Discrete
inhA	Continuous
katG	Continuous
kasA	Continuous
ndh	Continuous
aphC	Continuous
MDR Resistant	Discrete
Sample source/Country	Discrete
Clade	Discrete

12.3 Specific Aims, Hypotheses, Models

The outcome of interest here is antibiotic resistance in M. tuberculosis. Logistic regression is used as the central model. As the data reflects a set of isolates collected across several countries the use of random effects is an option. Possible heterogeneity is also an issue. Cluster analysis is also of interest by country and other groupings, for example SCG grouping. ROC curves are used to assess the overall fit of the logistic regression model. The logistic regression models are examined for stability across various subgroups including previous infection and treatment groups.

The goal is to model resistance and assess parameters (expressible as odds ratios) in the statistical models. Frequentist confidence intervals and p-values are used as well as Bayesian marginal posterior credible regions and odds ratios, Bayes factors when appropriate. Note that each of the considered genes here has a biological interpretation in relation to the onset of antibiotic resistance. This is assumed known here and can be employed to justify prior selection reflecting whether the parameter in question is related to a gene acting as a promoter, enhancer, inhibitor, or is simply associated with antibiotic resistance in TB.

Questions to be addressed include the following:

1. Develop a logistic regression model for antibiotic resistance. Comment on its properties and its predictive aspects.

2. Discuss the selection of prior densities for various genes being expressed.

3. Using cluster based methods to examine clustering patterns among individuals and among variables of interest.

TABLE 12.2

Data

		inhA(0.82)		ndh(0.69)		ahpC(7.42)		kasA(0.06)		katG(16.3)	
		Hi	Lo	Hi	Lo	Hi	Lo	Hi	Lo	Hi	Lo
INH Resist	Yes	36	16	18	33	32	21	6	44	35	18
	No	44	44	22	27	8	39	34	16	5	42

TABLE 12.3

Correlation Matrix

Non-Resistant Cases				
	ahpC	inhA	kasA	ndh
inhA	0.775*			
kasA	-0.668*	-0.775*		
ndh	-0.120	-0.221	0.248*	
katG	0.763*	0.778*	-0.755	-0.179
Resistant Cases				
	ahpC	inhA	kasA	ndh
inhA	0.101			
kasA	-0.027	0.158		
ndh	0.146	-0.286	0.039	
katG	-0.127	-0.099	-0.209	0.162

12.4 Analysis and Interpretation

12.4.1 Data Analysis

The data reflect simulated values based on Hazbon et al., (2006). The data set is available in supplementary materials. The continuous gene expression variables are rescaled here to provide a standardized expression index [0 - 100] and also given a median threshold and re-expressed as categorical variables [High, Low] to improve the interpretation and fit of the logistic regression. This also allows for two at a time contingency table analysis, comparing each gene expression variable to the INH resistant outcome. It is possible for these results to differ from the overall model if the explanatory variables are correlated. These are shown in Table 12.2.

All of these are individually significant at the 0.01 level except for the ndh variable. A correlation matrix for the gene expression variables is given in Table 12.3 (starred indicates significant). The correlation structures differ among resistant and non-resistant cases.

TABLE 12.4

Logistic Regression Summary

Predictor	Coefficient	Std Error	Z	p-value	OR	OR_l	OR_u
Constant	-1.96	0.97	-2.04	.042	–	–	–
ahpC	0.45	0.65	0.69	0.48	1.57	0.44	5.66
inhA	2.02	0.70	2.87	.004	7.53	1.89	29.95
kasA	-1.17	0.71	-1.65	.099	0.31	0.08	1.25
ndh	-0.20	.56	-0.36	0.72	0.82	0.27	2.47
katG	0.88	0.78	1.13	0.26	2.41	0.52	11.11

12.4.2 Likelihood Function

Assuming a binomial error distribution here as well as a logistic regression structure we have as our initial overall likelihood function:

$$L(\boldsymbol{\theta} \mid \mathbf{y}) = c \cdot \prod_{i=1}^{n} \left[\frac{\exp(\sum_{j=0}^{k} \theta_j x_{ij})}{1 + \exp(\sum_{j=0}^{k} \theta_j x_{ij})} \right]^{y_i} \left[1 - \frac{\exp(\sum_{j=0}^{k} \theta_j x_{ij})}{1 + \exp(\sum_{j=0}^{k} \theta_j x_{ij})} \right]^{1-y_i}$$

Here we use this basic model to obtain the ROC curve, odds ratio estimates and extend this slightly to a random effects logistic model.

12.4.3 Likelihood Frequentist Analysis

The model parameters are estimated throughout using *m.l.e.* based estimators and standard errors from the large sample (Cramer-Rao) bound variance-covariance matrix. P-values reported for hypotheses in question use standard pivotal quantities based also on the *m.l.e.*. Based on the assumed likelihood and observed data we have as our estimates and 95% confidence regions (Table 12.4).

Logistic regression modeling was conducted using the five genes. The results of this analysis confirm mutations in ndh was not associated with INH resistance (p-value = 0.82). In order to address the potential clustering of samples within a country, the final logistic regression model was rerun, allowing for correlation within the country. There was no significant change.

12.4.4 Likelihood Bayesian Analysis

Within the context of the logistic regression model the parameters of interest reflect the odds ratio linking the particular variable in question with the TB response, assuming all other variables are held constant. There are some expectations here regarding the parameters in question. Some of the gene expression variables are thought to have some effect on the response. The correlation structure among the gene expressions is not known and is not assumed here at the level of prior.

TABLE 12.5
Prior Densities

Parameter	Prior Density
inhA	$\log(\theta_1) \sim N(2, 10)$
ndh	$\log(\theta_2) \sim N(1, 10)$
kasA	$\log(\theta_3) \sim N(1, 10)$
katG	$\log(\theta_4) \sim N(2, 10)$
ahpc	$\log(\theta_5) \sim N(.5, 10)$

Given this assumed information, independent Normal priors are assumed for the parameters on a log scale and the parameters rescaled in the likelihood function. As the logistic regression model is a nonlinear model we rescale the parameters first and then assume the priors on this scale. No formal change of variable is required. We therefore expect for example the inhA variable to be associated with the outcome and therefore having an odds ratio greater than one. A reasonable assumption for a prior would seem $\log(\theta_j) \sim N(2, 0.0001)$ which reflects this belief and remains somewhat diffuse. The log scale is useful here to remove any restriction on the odds ratio values, which is good practice in most modeling situations. We can do this for all variables being considered. Note that we can also assume a mean of 1 to investigate a belief in no association. Table 12.5 shows the assumed set of priors reflecting existing theory.

Note that a Student-t prior density considered in the same manner is sometimes useful in the logistic regression here as well. We assume here that the diffuse normal prior densities (N(0,0.0001)) are independent throughout and no information regarding possible correlations among the gene expressions. This can be added by considering multivariate-t or Normal densities with sutable correlation matrices. The associated 95% HPD credible regions are given in Table 12.6 and compared to the relevant *m.l.e.* 95% confidence regions under two sets of priors; non-informative and those reflecting existing theory.

12.4.5   Analysis/Integration/Comparisons

The careful development of models that reflect associations between the presence of antibiotic resistance and specific genetic structures, here in the parasite itself, is both important and timely. Such results require replication and study in various sub-populations to be viewed as supported theory.

Here the application of logistic regression gives a useful model giving approximately 80% concordance of predicted onset with actual values. The genes in question are known to some extent in the sense of restricting or promoting resistance. The analysis here is limited and there are probably other genes of relevance as typically the presence of strengthened resistance in a phenotype

TABLE 12.6
Ninety Five Percent Confidence and Credible Intervals

Prior Assumption	ndh	inhA	kasA	ahpC	katG
Non-Informative	(0.31, 2.31)	(1.78, 28,11)	(0.068, 1.17)	(0.63, 5.31)	(0.41, 10.87)
Normal	(0.37, 2.23)	(2.7, 27.6)	(0.059, 1.15)	(0.75, 5.01)	(0.23, 9.64)
m.l.e. No prior	(0.27, 2.47)	(1.89, 29.93)	(0.08, 1.25)	(0.66, 5.66)	(0.52, 11.11)

implies a weakness in some other aspect of overall fitness as an organism has a finite amount of energy.

The fitted frequentist model here shows inhA as significant and kasA trending towards significance. The marginal Bayesian results for each parameter (odds ratio) does not alter the result, though shorter credible intervals are observed versus the confidence intervals. This often occurs and it should be mentioned that the integration or averaging out of four parameters can have a narrowing effect on these intervals. Note that once these types of results are validated, the use of informative priors becomes much more acceptable as theory reflects existing knowledge.

12.5   Suggested Exercises

1. TB often does not exist in an isolated state. Co-morbidities such as HIV, drug addiction, poverty and socio-economic factors are often present and difficult to measure accurately. Discuss how to model these effects in an international sample of TB isolates if using a linear model. What type of prior densities would be required in such an analysis?

2. Use Bayes factors to compare the various possible logistic regression models for resistance, where we choose different sets of explanatory variables that are useful in the logistic regression model above. Do we see a pattern in the Bayes factor values as we add information (each variable) or do we simply see random or Brownian motion?

3. Consider modeling the relationship between TB genetics and the genetics of the human immune system. Given the long time frame of human infection by TB, this is almost certainly present on some scale. Assuming that respective sets of say 8 genes for each are identified as relevant, discuss how to apply frequentist multivariate methods such as canonical correlation to assess the connections between the genetic factors. If a Bayesian perspective through the application of a linear model is to be applied (i) what would prior densities look like here? (ii) how might random effects considerations enter the analysis?

4. Differences in the data overall and within subgroups can be assessed with regard to inferences for a given parameter θ_j by measuring the change in the marginal posterior $p(\theta_j|data)$ for a given prior density. One measure that can be used for this is a version of the entropy measure; $E_{i,k} = \int p_i(\theta_j|data) \log(p_i(\theta_j|data)/p_k(\theta_j|data))d\theta_j$ where k represents the overall data and i the subgroup of interest. If the $E_{i,k}$ values are fairly constant the analysis regarding θ_j can

be seen as stable in terms of the marginal posteriors for θ_j. Use the entropy measure based on marginal priors to examine the stability of inference by country. Is there evidence of Simpson's paradox (Bayesian or frequentist perspectives) across country?

12.6   Bibliography

[1] Brimacombe M., Hazbon M., Motiwala A.S., Alland D. (2007). Antibiotic Resistance and Single Nucleotide Polymorphism Cluster Grouping Type in a Multi-national Sample of Mycobacterium Tuberculosis Isolates. *Antimicrob Agents Chemother*. 2007 Nov;51(11):4157–9. Epub 2007 Sep 10. PMID: 17846140.

[2] Cohn D.L., Bustreo F., Raviglione M.C. (1997). Drug-Resistant Tuberculosis: Review of the Worldwide Situation and the WHO/IUATLD Global Surveillance Project. *Clinical Infectious Diseases* 1997; 24(Suppl 1):S121–30.

[3] Hazbón M.H., Bobadilla del Valle M., Guerrero M.I., Varma-Basil M., Filliol I., Cavatore M., Colangeli R., Billman-Jacobe H., Lavender C., Fyfe J., García-García L., Davidow A., Brimacombe M., Leon C.I., Porras T., Bose M., Chaves F., Eisenach K.D., Sifuentes-Osornio J., Ponce de León A., Cave M.D., Alland D. (2005). Role of embB codon 306 Mutations in Mycobacterium Tuberculosis Revisited: A Novel Association with Broad Drug Resistance and IS6110 Clustering Rather Than Ethambutol Resistance. *Antimicrob Agents Chemother*. 2005 Sep;49(9):3794–802. Erratum in: *Antimicrob Agents Chemother*. 2005 Nov;49(11):4818. PMID: 16127055.

[4] Hazbón M.H., Brimacombe M., Bobadilla del Valle M., Cavatore M., Guerrero M.I., Varma-Basil M., Billman-Jacobe H., Lavender C., Fyfe J., García-García L., León C.I., Bose M., Chaves F., Murray M., Eisenach K.D., Sifuentes-Osornio J., Cave M.D., Ponce de León A., Alland D. (2006). Population Genetics Study of Isoniazid Resistance Mutations and Evolution of Multi-Drug Resistant Mycobacterium tuberculosis. *Antimicrob Agents Chemother*. 2006 Aug; 50(8):2640–9. PMID: 16870753.

Index

Printed and bound by PG in the USA